Hardy, Randall W.

China's oil future

DATE			

China's Oil Future

China's Oil Future:
A Case of Modest Expectations
Randall W. Hardy

Much has been written recently about China's emergence as a potential oil power. Comparisons have ranged from those that picture China as another Middle East, with a Middle East-like impact on future global oil markets, to more modest images of an oil-producing nation that can meet its rapidly expanding internal needs through the late 1980s and still have some oil for export to its neighbor, Japan. Yet to fulfill even the latter prediction, the People's Republic of China will have to surmount a series of substantial political and technical obstacles. This book identifies those constraints, assesses the likelihood of China's overcoming them, examines the incentives for increasing Chinese petroleum exports, and analyzes the role such exports could play in Peking's foreign policy.

Randall W. Hardy is assistant to the regional representative, Region X (Seattle), of the U.S. Department of Energy. He was previously special assistant to the administrator, Federal Energy Administration, in the areas of energy conservation and resource development, oil price and allocation regulations, and domestic and international energy policy.

China's Oil Future: A Case of Modest Expectations

Randall W. Hardy

Westview Press • Boulder, Colorado

Dawson • Folkestone, England

This volume is included in Westview's Special Studies on China and East Asia.

Any views, conclusions, or recommendations contained in this report are attributable to the author and should not be interpreted as necessarily reflecting those of the Department of Energy or the federal government.

Published in 1978 in the United States of America by
Westview Press, Inc.
5500 Central Avenue
Boulder, Colorado 80301
Frederick A. Praeger, Publisher

Published in 1978 in Great Britain by
Wm. Dawson and Sons, Ltd.
Cannon House
Folkestone
Kent CT19 5EE

Library of Congress No: 77-27555
ISBN (U.S.): 0-89158-156-1
ISBN (U.K.): 07129-0871-4

Printed and bound in the United States of America

To
Janie
and I.B., B.B., and Poke

Contents

Tables

Preface

Much has been recently written about China's emergence as a potential world oil power. Comparisons have ranged from picturing China as another Middle East, with a similar impact on future global oil markets, to the more modest image of an oil-producing nation which can meet its rapidly increasing internal needs through the 1980s and still have significant oil for export to a nearby Japan.

Yet even to fulfill the latter prediction, the People's Republic of China (PRC) will have to surmount a series of substantial political and technical obstacles. This paper will endeavor to identify those constraints, to assess the likelihood of China's overcoming them, to examine the incentives for expanding Chinese petroleum exports, and to analyze the role such exports could play in Peking's foreign policy.

The study itself is the product of some two years of research. Its quality has been improved considerably through criticism by my colleagues, both inside and outside the federal government. The responsibility for any errors, omissions, and other defects is mine alone.

Of the many people who assisted me in this endeavor, two deserve special mention: Mel Conant of International Energy, for the inspiration and support to undertake the effort initially, and Sam Tuthill, now an Iowa utility executive, for taking the time to explain the basics of geology to a struggling political scientist.

I also wish to thank Linda Chan and Barb Shurin, whose superb typing efforts enabled me to meet what seemed at the time to be impossible deadlines.

And I feel, as usual, forever indebted to my wife, whose suggestions helped me structure the manuscript and whose patience sustained us both through two years of endlessly discussing what came to be known as China's "great spurt forward."

Summary

Perhaps the best way to examine PRC oil prospects is to postulate future production levels and then to identify barriers to their accomplishment. Using that approach, this paper concludes the following major constraints must be overcome to reach the much-quoted output figures of 4 million barrels per day (bpd) in 1980 and 8 million barrels per day in 1990.

• Continued and even closer Sino-Japanese cooperation will be necessary throughout the entire fifteen-year period (1975 to 1990), as Japanese technology will provide key assistance to China's petroleum development, and a growing Japanese oil market will be the main recipient of PRC exports. Continued close relations seem likely, but divergent interests in Taiwan, South Korea, and the Senkaku Islands; conflicting perceptions of proper Chinese-Japanese roles in East Asia; and a history of mutual antagonism could all undermine the current rapprochement.

• The PRC must maintain and probably expand its present "prodevelopment" policies for the next fifteen years, as political stability will be a prerequisite for significant oil industry growth. This would require foregoing future mass mobilization campaigns and would assume that Chou En-lai–type pragmatists continue to dominate China's post-Mao leadership.

• Increasing percentages of oil-related investment will be necessary from a relatively finite capital pool, especially if most production increases after 1980 come from offshore or interior fields. Yet these capital increases must occur during the same period when coal, steel, and agriculture are making similar demands for greater investment shares.

• Sufficient reserves must exist to support projected production increases. Limited geologic information and the remote location of many Chinese basins, however, could limit the amount of readily recoverable reserves.

• Present and future bottleneck problems in those industries which support oil must be resolved. This means overcoming recent lags in coal and steel output, as well as avoiding anticipated overloads of China's transportation systems.

• The PRC must maintain an adequate level of technological expertise to produce sufficient geologists, geophysicists, and production engineers to run a sophisticated modern oil industry. This will require skill training undiluted by post–Cultural Revolution educational reforms and will demand technicians capable of rapidly absorbing recent Western innovations.

It is important to recognize the *simultaneous nature* of these development constraints. Chances are probably 80 or 90 percent that each potential difficulty, when considered by itself, can be positively resolved. However, the composite probability of reaching 8 million bpd in 1990, when such problems are taken together, is probably well under 50 percent. As a result, these supply constraints and rising domestic demand should significantly restrict Chinese oil exports for the foreseeable future.

Depicting China as a potential oil giant is therefore extremely misleading. The PRC should continue to use its oil "weapon" to influence Japan—primarily by discouraging its participation in Siberian projects—and to score political points with the Third World. But China's limited export prospects, quality problems associated with PRC crude, and competition from Indonesian Minas crude will likely prevent oil's emergence as a primary Chinese foreign policy tool. It will be a complicating but not controlling factor in the future geopolitics of Asia.

1
How Much Oil and Where

China is currently producing roughly 1.8 million bpd of crude oil, with annual production having increased by over 20 percent since 1970. Much of the available literature typically projects future Chinese production as 4 million bpd, with 1 million bpd for export, by 1980.[1] This figure has been buttressed by frequent PRC discussion of the 1 million bpd 1980 export level with Japanese officials. Longer range projections are usually somewhat more conservative for 1980 but still indicate production rates of 8 million bpd by 1988 or 1990.[2]

For comparative purposes, recent production and export figures are shown in Table 1.[3] One trend seems to be emerging: for those years when China has exported significant quantities, its exports have comprised only 10 to 15 percent of total production. Even with a 4 million bpd production level in 1980, it thus appears unlikely that exports then would exceed 600,000 bpd.

In addition, a 1975 Central Intelligence Agency (CIA) energy balance study suggests that the most likely PRC production and export figures for 1980 are 3.0 to 3.4 million bpd and 540,000 to 660,000 bpd,[4] caused primarily by a growth in total domestic energy demand of 8.4 to 11.4

Table 1

Domestic Oil Use and Export Levels
(In Thousand Barrels Per Day)

	Year of Production					
	1972	1973	1974	1975	1976	1977[a]
Domestic Supply	860	1,033	1,191	1,248	1,475	1,641
Japan	---	20	80	162	122	132
Hong Kong (all products)	---	1	5	13	13	14
Philippines	---	---	3	13	10	18
Thailand (mostly products)	---	---	1	4	10[a]	5
Vietnam/Cambodia (all products)	---	10	10	10	---	---
North Korea	---	16	10	20	20[a]	20
Rumania	---	---	---	10	30[a]	---
Total Production:	860	1,080	1,300	1,480	1,680	1,830
Total Exports:	0	47	109	232	205	189

[a]Estimated

percent annually. The CIA believes 1 million bpd of oil exports by 1980 can only be achieved at the expense of considerable economic growth,[5] and they dismiss Chinese mention of such a figure by citing Peking's desire to lure Japan away from participation in Siberian development, the natural tendency of central planners to overstate expected results, and the vested interest of those in power in demonstrating the success of their policies—for example, that increased oil exports will bring larger capital goods imports to accelerate PRC development.

To reach any supportable conclusions about China's potential, however, production rates and reserve estimates for individual fields and geologic basins must be analyzed. The chart in Appendix A shows that the Ta-ch'ing field in northern Manchuria and the Sheng-li and Ta-kang fields southeast of Peking near the coast are the PRC's main present producing areas. Together they accounted for roughly 1.3 million bpd of China's 1976 production total of 1.7 million bpd. Ta-ch'ing was discovered in 1959 and has been the PRC's main producing field since the mid-1960s, accounting for 40 to 50 percent of Chinese production. Sheng-li and Ta-kang, on the other hand, have only become major producers in recent years but are the most rapidly growing of all known PRC fields. Ta-kang is of particular importance, as it lies adjacent to the Po Hai, a gulf north of the Yellow Sea whose offshore potential the Chinese are now vigorously exploring.

If calculating Chinese production figures is difficult, estimating the resources of different basins with adequate precision is close to impossible. Although such figures are a necessary prerequisite to assessing China's ultimate potential, they should be regarded with considerable skepticism for a number of reasons. First, as only generalized stratigraphic data are available, very little is known about the distribution and characteristics of the deep source beds in China's numerous basins. Since such rock sequences are the basic geologic "building blocks" from which oil is created,[6] this factor alone could cause an error in resource estimation of 100 percent or more.

Next, few data are available on the amount of trapped pore space, porosity distribution, oil saturation, or age of geologic formation, so computing net acre-feet of trapped pore space is not a reliable alternate method of resource estimation.

Finally, there is no adequate measure of hydrocarbon loss from the system through time. This consideration includes the questions of how much oil is lost both inside and outside a particular structural system. The result is that even in the United States, with usually large amounts of information, major oil companies have lost hundreds of millions of dollars drilling dry holes in supposedly promising areas (e.g., the Destin Anticline off Florida). The corollary is that for China, an area for which only generalized geologic data exist, credible resource estimates have been replaced by rather rudimentary "best guesses."

With an awareness of these basic limitations, it is possible to examine what is known about resources in the PRC's main basins, whose locations are shown in Appendix B. With a few exceptions, most of China's potential oil areas are in continental lacustrine basins.[7] In eastern China these basins are characterized by a large number of highly fractured, possibly stratigraphic, traps. The U.S. Geological Survey (USGS) suggests that the depositional environment and reservoir conditions of these main producing fields are quite analogous to the Uinta basin in northern Utah.[8] The principal result of such a geologic setting is great variability in reservoir distribution and in porosity and permeability within each basin. Rather than large reservoirs which are easily located and exploited, Ta-ch'ing, Sheng-li, and Ta-kang all appear to have oil in an unpredictable series of smaller pockets at various depths.

A closer look at the drilling patterns in each of China's

major fields also seems to confirm the difficulties associated with their development. For example, observer reports indicate that the Ta-ch'ing complex consists of some 4,000 wells spread out over several hundred square miles.[9] Dividing that field's 1975 output (780,000 bpd) by the number of wells yields an average production rate of roughly 200 bpd per well, a figure generally consistent with other estimates of Ta-ch'ing's daily per-well output.[10] But whether 100 or 300 bpd, it seems clear that China's largest field is characterized by many wells with rather modest pumping rates—at least by Middle East standards—scattered over a wide geographic area. This pattern of complex geology and consequent exploration and production problems also repeats itself at both Sheng-li and Ta-kang.[11] Although discovered shortly after Ta-ch'ing, Sheng-li production over fourteen years has only reached 300,000 bpd, whereas Ta-ch'ing output over sixteen years has equaled 800,000 bpd. Geologic complexities no doubt are a primary factor in this slower production rate, but severe weather, flooding by the Yellow River, and diversion of most skilled manpower to Ta-ch'ing were also contributing elements.

Finally, China's second-order fields appear to face similar circumstances.[12] P'an-shan, located just north of the Po Hai, reflects characteristics of Sheng-li and Ta-kang; Fu-yu, located just south of Ta-ch'ing, has that field's general structure; Ch'ien-chiang, located in central Hupeh Province, has no nearby analog but has not experienced any dramatic growth in its eight-year existence. Regardless of its exploitation difficulties, however, Ch'ien-chiang will likely continue to receive attention as its proximity to oil-poor south China creates a strong incentive to tap reservoirs in spite of their marginal output.

If correct, the above conditions create several difficulties for the Chinese. First, they must sink many wells with relatively close spacing and at different depths to maximize their exploitation of various reservoirs.[13] Even with this high well density for a given area, their probability of finding dry holes is still considerable. Second, the percentage of oil ultimately recoverable from such fields is likely to be low by world standards. In addition, oil produced from these basins is usually characterized by a high wax content (20 to 30 percent) and a high pour point (95°+ F), requiring elaborate heating facilities to extract and transport it. Such circumstances might dictate a relatively high capital investment/unit of output ratio, and, if these structures extend offshore under the Po Hai,[14] a rather costly petroleum development process could result.

Despite indicators which may portend significant production problems, the lack of more complete information and the nature of development to date could also lead to different conclusions. Combined structural and stratigraphic traps are more difficult to map than simple structural ones, but they do not necessarily contain smaller quantities of oil.[15] The exploration of oil in stratigraphic traps, however, generally requires greater technological sophistication. Since thoroughness of exploration proceeds in tandem with the level of petroleum technology, Chinese acquisition of better seismic equipment and mastery of more advanced exploratory techniques may make possible larger future discoveries than now appear likely. This possibility is heightened by the fact that most Chinese drilling has been at depths less than 10,000 feet.[16] Deeper drilling could yield discoveries large enough to change the development picture, but at this point China's geologic problems seem to overshadow its potential for new, easily exploitable discoveries.

The recent production history of Ta-ch'ing itself tends to confirm such problems. After average annual output gains of 19 percent from 1965 to 1975, Ta-ch'ing's 1976 production increased by only 8.7 percent.[17] This slowdown was accompanied by persistent reports that Ta-ch'ing's production may be on the verge of a major decline.[18] The reasons cited were:

1. Graphs at one "model" field showing production increases were only continued through tripling the amount of water injection.
2. Drilling of new wells and of exploratory wells on fringes of Ta-ch'ing appeared to be completely halted.
3. Many wells appeared abandoned or had been converted to "stripper" status.
4. Only small amounts of new housing were planned, indicating an expectation that population would decrease as workers moved to more promising fields.

Although geologic factors were in part responsible for these changes, there appear to have been at least two other reasons. First, Ta-ch'ing has been water-flooded since its inception,[19] often without much attention to the lateral and vertical changes in reservoir porosity and permeability throughout each field. The result has been frequent coning and bypassing of productive pockets, thus physically damaging and shortening the productive life of individual fields. Second, the Chinese have recently begun switching from simple line drive water-flooding to a more complicated five-spot or nine-spot water-flood technique.[20] This approach greatly decreases the number of injection wells required to supply a given quantity of extraction wells, thereby increasing sweep efficiency and the amount of oil ultimately recovered from a particular reservoir. A corollary effect to greater area coverage is a longer extraction period, with a concomitant slowing in the rate of increase of overall field production. Since the Chinese reportedly intend to convert much of Ta-ch'ing to this production method, its implementation also accounts for some of that field's output slowdown.

Not surprisingly, these geologic and production uncertainties have produced a wide range of reserve assessments.

The table in Appendix C shows overall reserve estimates from 20 to 95 billion barrels (BB). A large number of "informed sources" seem to settle on the 50 to 80 billion barrel range for total onshore and offshore recoverable reserves.[21] Yet the amount of recoverable PRC reserves is not nearly as important as its distribution. Appendix C seems to indicate the following pattern of location for Chinese reserves, regardless of the total estimated reserve base:

Area	Percentage of Total PRC Reserves
Offshore Fields	20 - 40%; Midpoint = 30%
East China Fields	15 - 25%; Midpoint = 20%
Interior Fields	45 - 55%; Midpoint = 50%

Despite some difference, it can be concluded that the PRC is currently producing most of its oil from the region with *least* potential reserves. This condition is likely to continue until at least 1980, given the long lead times involved in offshore development. Assuming that China reaches a total production level of 3.0 to 3.4 million bpd in 1980, it will have produced about 7 billion barrels to that date. From 1980 to 1990, just to sustain a 3.5 million bpd production level, the PRC will use another 12 to 13 billion barrels.[22] Even with the most optimistic overall reserve estimates (100 BB) distribution ratios give eastern China maximum reserves of only 20 billion barrels. It appears, therefore, that China's eastern fields could be exhausted by 1990 unless new sources are developed and that nearly all post-1980 production *over* 4 million bpd must come from development of either offshore or interior fields.

An alternative means of viewing the problem would be to compute the probable reserve base of *eastern* China fields necessary to sustain annual production increases of 15 percent from 1976 to 1980 and 10 percent thereafter. Most major oil companies use a reserve/production ratio (R/P) of 30 to calculate sustainable production rates; e.g., a production rate of 50 million tons per year (1 million bpd) would require a reserve base of 1.5 billion tons, or 11 billion barrels. The tables in Appendix D show the total reserve base and annual discoveries necessary to maintain R/Ps of 30 and 15 between now and 1983. Using a maximum initial reserve figure of 20 billion barrels for eastern China, just to maintain an R/P of 30 (the world average), China would need to discover a Ta-ch'ing-size oil field every year starting in 1977. To maintain an R/P of only 15, the same discovery rate would be required, but commencing in 1980.[23] In both cases the conclusion seems clear: post-1980 Chinese production increases probably will come from sources other than eastern China.

One caution should be added in projecting these various reserve and production levels. Most existing estimates assume that the PRC has not discovered any large unpublicized oil reserves. In the past, however, the Chinese have frequently withheld information on new fields until they were fairly well developed.[24] Although it is possible that this practice will continue, Peking's 1974 decision to expand exports has been accompanied by a marked increase in oil data released to the outside world. Since China has a considerable stake in convincing potential buyers (such as Japan) of its promise as a long-term supplier, it seems unlikely that the PRC would have concealed news of large discoveries which could significantly augment its output.

Given this rather pessimistic long-term outlook for eastern China, what kind of petroleum development strategy have the Chinese pursued to date, and what approach are they likely to adopt in the future? From 1950 to 1960, China relied heavily on Soviet equipment and assistance to survey and develop potential fields. The result was that petroleum development was concentrated in northwest China, where the only known fields existed—an area also advantageous to the Soviets from a logistical and strategic viewpoint. With the discovery of Ta-ch'ing (1959), Sheng-li (1962), and Ta-kang (1964), the pattern gradually shifted to northeast China. This shift was facilitated by having the discoveries relatively near major Chinese population and industrial centers. The withdrawal of Soviet technicians and support in 1960, although significantly hampering China's immediate rate of oil growth, did stimulate indigenous Chinese efforts and eventually aided the PRC's push toward self-reliance in the petroleum industry.

Another characteristic of this development pattern was its sequential nature. Not only did the actual location of discoveries themselves dictate exploitation first of the Northwest and then of the Northeast, but limited manpower capable of drilling and production operations also prevented simultaneous development of these areas. By concentrating resources in one geographic region at a time, the Chinese maximized training and material resources for production while allowing more time for frontier area exploration.

Despite the Soviet withdrawal in 1960, Chinese reliance on Russian techniques continued until the early 1970s. It was manifested by the widespread use of Soviet and Rumanian equipment and by following the USSR development strategy of exploiting the large, easily accessible structures first. Up to now, this relatively straightforward approach has worked well. The Chinese, despite more recent use of multiple completion and secondary recovery equipment, still largely drill on domes or other high probability targets and count on striking oil at relatively shallow depths.[25] In addition, past oil activity has concentrated on geographically accessible areas. Remote fields, even those with great potential, have purposely been left for subsequent exploration due to the magnitude of transportation and infrastructure problems associated with their development.

Since 1973 onshore Chinese exploration and production activities have intensified,[26] but the basic pattern has not changed. A new element has been China's purchase of Western oil equipment and the emergence of offshore developments. With purchases from the United States totaling over $100 million,[27] China has made a definite commitment to using foreign equipment for petroleum development. Purchases cover a variety of areas, from seismic exploration equipment through well-logging apparatus to well-blowout preventers. But procurement trends indicate that most drilling and production purchases are apparently intended for onshore development, while exploration buys are aimed largely at offshore resources. There will probably be continued emphasis on expanding existing onshore fields between now and 1980. Meanwhile, offshore exploration activities, in the Po Hai and elsewhere on the continental shelf, are designed to help the Chinese proceed with gradual, more cost-effective water-borne production in the period after 1980.

This tendency toward expanding present fields while continuing offshore exploration is reinforced by general marketing considerations. China already has the pipelines, port facilities, and refineries in place to handle oil from Ta-ch'ing, Sheng-li, and Ta-kang, both for export to Japan and for distribution to her own industrial centers. Moreover, there is typically an inclination to explore close to currently producing wells, almost regardless of geological indicators, rather than undertake real wildcatting in unexplored basins far from potential markets. Infrastructure investment requirements, therefore, probably dictate China's development dynamic as much as geological considerations.

This continued emphasis on eastern fields now and offshore later, while seemingly practical, is not without problems. As mentioned before, the highly differentiated geologic environment of eastern China could make further development increasingly costly as more wells with low pump rates are required to boost overall production. Such factors will likely be exacerbated as the Chinese are forced to drill deeper and problems such as overpressuring and lost circulation occur.[28] In the long term, offshore areas will be hard pressed to make rapid additions to total production because of their greater capital requirements and the Chinese belief in self-development of resources.

2
Marketing and Distribution

Given that the Chinese have significant quantities of oil, there are still major variables affecting its delivery to both external and internal markets. Refineries of various sizes, locations, and capabilities are necessary to convert crude into product forms appropriate for end-use consumption. Pipelines, railroads, port facilities, tankers, and other transportation mechanisms are required to carry petroleum from the production site to the point of eventual use. Quality and price must also be such that the PRC can successfully market its oil abroad.

Of these variables, refineries appear easiest to assess. As detailed in Appendix E, total PRC capacity equaled roughly 1.3 million bpd in 1975, compared with an overall crude production level of 1.5 million bpd. To date, refinery enlargement has kept pace with production growth. Although similar future expansion will require greater amounts of capital, the PRC has planning flexibility sufficient to obviate a direct correlation between rising refinery capacity and well output. Most importantly, China can burn some crude directly in power plants and other facilities.[29] It also can increase product outputs through technical modernization of current refineries rather than more costly construction of new ones.[30] In addition, the asymmetry between foreign and domestic demand for both crude and refined products, plus the PRC's ability to control its own supply investment and aggregate demand, give China a wide range of acceptable product output levels. If capacity threatens to lag production growth significantly for a particular year, the excess oil can probably be exported or simply shut in, and investment planning targets then readjusted as necessary. As shown by the refining capacity and utilization chart in Appendix F, the Chinese have managed a tight supply situation in recent years by simply burning more crude in power plants and increasing the rate of refinery utilization.

Peking has also apparently committed itself to a 15 percent annual refinery growth rate for the next five years.[31] This goal will be required to meet both overall demand growth and specific requirements of the petrochemical sector.[32] But with no obvious bottlenecks to construction and such flexible output margins, refinery construction alone does not appear to be a significant obstacle to production increases in general or export expansion in particular.

The same conclusion appears valid for PRC port facilities. While Table 2 indicates that known port size[33] presently constrains dramatic increases of oil exports, a vigorous expansion effort is underway. Recent construction has already led to substantial improvements in PRC port capacity. From 1973 to 1975, the Chinese built forty new berths capable of handling general cargo and oil vessels of over 10,000 dead weight tons (dwt).[34] As a signal of future plans, China has also expanded its merchant fleet by the purchase in late 1975 of three 90,000 to 100,000 dwt tankers.[35] With combined growth in both port facilities and PRC-owned tankers, China should overcome current transshipment constraints by 1978-79. Her ports still would not accommodate the supertankers (200,000-plus dwt) needed, at current prices, for exports to the United States, but they should be capable of handling any foreseeable near-term export increases to Japan.[36]

Table 2

Port Location and Capacity

Port	Location	1976 Capacity Remarks
Dairen (Ta-lien or Lü-ta)	Tip of Liao-tung Peninsula, east end of Po Hai.	100,000 dwt with 1 or more 50,000 dwt berths; additional 100,000 dwt berths believed under construction; main PRC tanker port. Separate harbor at old Lü-ta, 8 mi. to SW of principal harbor, has 30,000 dwt capacity; used primarily for shipments to Shanghai.
Ch'in-huang-tao	East of Peking, northwest side of Po Hai.	50,000 dwt; second wharf of 70,000 dwt capacity reportedly under construction. Also several 35,000 and 20,000 dwt berths.
Chan-chiang	South China, just north of Hainan Island.	50,000 dwt with possible 70,000 dwt capacity; used mainly for off-loading crude imports for South China.
Tsingtao (Huang-tao)	East Shantung, northwest side of Yellow Sea.	Transshipment point for Sheng-li/I-tu crude; has at least one 50,000 to 70,000 dwt berth.
Ports of Unknown But Probably Smaller Oil Handling Capacity:		
Shanghai	Center of east China coast.	Frequent berth for offshore drilling platforms. May have one or more 50,000 dwt berths.
Hsing-kang	Tientsin Harbor, southeast of Peking.	Possible transshipment point for Sheng-li crude. Reportedly has two 25,000 dwt berths.
Whampoa	Canton Harbor	Possible transshipment point for Kwan-tung (Nanhai County) oil.

Inland transportation is another potential bottleneck that appears to have been cleared for the moment. Prior to 1970, the PRC relied mostly on railroad tank cars and coastal tankers to move crude oil. A concerted pipeline construction effort has now largely replaced this former means of transport, providing faster, more efficient delivery of oil to principal distribution points. Most of the long-distance pipelines have diameters of 8 to 24 inches, with the latter size being imported from Japan.[37] Although several PRC pipelines are well situated to facilitate exports (e.g.,

Ta-ch'ing to Ch'in-huang-tao, Ta-ch'ing to Dairen, and Sheng-li to Tsingtao), the bulk of China's effort seems to be domestically oriented. Table 3 illustrates the scope of recent activity—from roughly 600 pipeline miles in 1970 and 1971 to nearly 3,000 miles completed and 1,300 miles under construction in mid-1976.[38]

Table 3

Operational Oil Pipelines

Pipeline	Length (Miles)	Remarks
I. Known To Be Complete:		
1. Yu-men/Lan-chou*	450	16-inch diameter
2. K'o-la-ma-i/Tu-shan-tzu (2)*	90 ea.	One 16-in., one 24-in. diam.
3. Ta-ch'ing/Ch'in-huang-tao	600	24-inch diameter
4. Ta-ch'ing/Dairen	620	24-inch diameter
5. Lüng-nu-ssu/Ch'ung-ch'ing	190	----
6. Ching-men/Ch'ien-chiang	60	----
7. T'ieh-ling/An-shan	120	----
8. K'o-la-ma-i/Wu-lu-mu-ch'i	190	----
9. Lin-i/Chi-nan	60	----
10. Sheng-li/Tsingtao	120	----
11. P'an-shan/Chin-hsi	60	----
12. Ch'in-huang-tao/Fang-shan (near Peking)	190	24-inch diameter
13. Chan-chiang/Mao-ming	90	Reportedly 39-inch diameter
Total:	2,930	
II. Believed Still Under Construction:		
1. Ta-kang/Fang-shan	190	----
2. Lin-i/Nan-ching	250	----
3. Lin-i/Po-hsing	60	----
4. Ko-erh-mu/La-sa	680	----
5. Sheng-li/Hsin-tien (2)	60 ea.	----
Total:	1,300	

*Pipelines in place as of 1970-1971.

As long as China's main fields are onshore in the eastern lowlands, overland transportation does not represent an insurmountable problem. But that sector's bottleneck potential could rise exponentially if the PRC must seek large amounts of oil from offshore or interior areas. In both regions standard construction difficulties would be magnified. Lack of trained civil engineers, adverse weather conditions, shortages of specialized steel, relatively high viscosity oil, and unfavorable terrain would all make the process more lengthy and expensive. Each area would also have a special set of limitations. Underwater pipelines, unless close to shore, would demand large applications of high-level technology which the Chinese probably lack. Interior pipelines, just by virtue of the distances involved,[39] would require massive, continuing resource investments to complete.

As a comparison, the Soviet Union has known since the mid-1960s of its huge, easily extractable oil and gas reserves in western Siberia. Despite the projected need for new sources to meet rising domestic and East European demand, the USSR has still been unable or unwilling to build pipelines and related facilities over the same distance (1,500 to 2,000 miles) necessary for the Chinese to get their oil to market. While there are clearly differences in climate and other factors, the experience of the USSR, with relatively greater capital investment resources than the PRC, illustrates the magnitude of development obstacles faced by relying on oil from western China, even if huge resources are discovered in that region.

Another consideration is the different motivations which underlie investments in China and the West. In Western nations economic forces (i.e., the profit motive) operate more or less on their own to overcome huge infrastructure problems. In the PRC, however, political forces must dictate large resource allocations without the natural incentive of a strong, consumer-based oil market. The latter approach requires that a substantial political consensus be maintained for five to ten years or longer, a process uniquely subject to midstream shifts in priorities, and one likely to require continuous demonstration of its near-term benefits for China—e.g., significant political leverage resulting from oil exports to specific countries.

A final near-term difficulty for China is oil quality. As Appendix G shows, Ta-ch'ing and Sheng-li crudes, while low in sulfur, are characterized by high pour points, high wax contents, large amounts of water and sediments, and high residual fractions.[40] Wax content has posed particular problems, from clogging pipelines to fouling refinery equipment. It was the main reason for the Philippines having to mix past imports of Chinese oil with Middle East crude before refining.[41] In addition, Appendix H shows that Ta-ch'ing crude has by far the highest residual percentage among world crude oils of comparable gravity (°API). This tends to limit the ease with which it can be refined into lighter products such as gasoline and heating oil, thereby placing such oil in direct competition with the heavier Indonesian Minas crude in Asian markets and restricting Chinese oil exports to fulfilling mostly heavy oil requirements of customers.

Conversion of importer refineries could solve or minimize all these problems, but the time and money necessary to achieve such capabilities, combined with Ta-ch'ing oil's current price of $13 per barrel, may hamper short-range export prospects. Indeed, two examples of marketability limitations have already appeared. First, Japanese oil imports from China decreased in 1976 and 1977 after two years of substantial growth. Second, China has been forced to sell Sheng-li crude to the Philippines for only $7 to $9 per barrel, and even at that discount price, Australia, New Zealand, and Japan have refused to buy. Whether all these quality problems will curtail China's post-1980 oil exports seemingly depends on PRC pricing policies and future petroleum demand in East Asia.

3
Offshore Development

China's recent push toward offshore development has excited particular interest in Western circles. One reason is a 1968 United Nations geophysical survey which concluded that the "Continental Shelf between Taiwan and Japan may be one of the most prolific oil and gas reservoirs in the world."[42] This was later buttressed by a Japanese survey which indicated significant oil-bearing sediments near the Senkaku Islands. Western companies such as Gulf, Amoco, and Shell have also been active in continental shelf areas near Taiwan and South Korea.[43] However, the exploratory drilling necessary to confirm this initial optimism in most East Asian seas remains to be done.

The geologic picture for offshore China is even more problematic than that for onshore basins, and about the only certainty is that neither foreigners nor the Chinese have yet acquired sufficient data to make reliable resource estimates. Such cautions, however, do not mean no data exist on this area's petroleum potential. Most geologists believe that sheer volumes of sediments indicate some promise, although they also agree that onshore extensions of nonmarine facies place definite constraints on the number and distribution of attractive drilling objectives, especially in the Yellow Sea and Po Hai regions.[44] The general area of speculation is shown in Appendix I, while Appendixes J and K are tabular and pictorial assessments of China's offshore potential by Jan-Olaf Willums, a Norwegian oceanographer who has taken the most comprehensive look to date at that area's promise.[45] Willums' "middle range" estimate of 29 billion barrels seems to show one reason why, even though the geology is complicated, the Chinese are pushing offshore.

But other PRC motives for offshore development are also clear. Besides the obvious lure of large reserves, offshore drilling helps China diversify from Ta-ch'ing, lessening the oil industry's vulnerability to possible Soviet actions. It likewise enables China to substantiate prior offshore territorial claims in areas somewhat removed from the Chinese coast. Even if actual drilling in many regions is not practical for several years, present declaration of Chinese intent backed by demonstrable activity in some waters assists the PRC in deterring exploration by potential rivals.

The potential for conflict among countries with rival claims should not be minimized. Although somewhat dated, the map in Appendix L shows the overlaps just among those nations and companies which are actively cooperating in offshore exploration. The PRC, in turn, claims nearly all this area for itself, based on the conviction that Taiwan and its surrounding waters are part of China, and on support for the international law principle of "national prolongation of land territory." This position works to Peking's advantage since the Chinese continental shelf slopes gradually eastward until separated from Japan and Okinawa by a deep trench. Even applying the superficially logical solution of equidistant division of the disputed territory (Japan's position) leaves unresolved questions about treatment of islands and archipelagoes. While the legal and territorial issues involved are beyond the scope of this study, it is important to realize that they will likely encourage the PRC to continue an aggressive offshore exploration program.[46]

Given these general incentives to move offshore, choice of the Po Hai for initial drilling was inevitable. The gulf contains China's two main oil ports and is near several major Chinese industrial areas. Its oil is thus ideally located for easy distribution either outside or inside China, an especially important domestic factor as transportation has been a traditional bottleneck for the Chinese petroleum industry. Equally important are Po Hai's shallow depths (between 50 and 125 feet) and calm weather, which minimizes the logistical problems associated with drilling other offshore regions. Furthermore, its promise was abetted by expanding discoveries at Ta-kang. Since many of that field's wells are located in coastal marshlands near the gulf,[47] a geologic link with offshore reserves seems probable, even without detailed analysis of the onshore/offshore basin structure.

Similar considerations of proximity, depth, and geologic promise will limit the amount of China's offshore activity far from Peking. Water depth in particular, due primarily to present levels of Chinese offshore technology, will play a large role in determination of future drilling patterns. With average depths of 180 and 220 feet, respectively, the Yellow and East China Seas appear feasible for development, but distinctly less attractive than the Po Hai. The South China Sea, with most of its floor more than 500 feet deep, is beyond PRC capability to exploit for the present. These conditions are reinforced by overall environmental factors. Appendix M displays wave frequency and height for China's offshore areas. Appendix N combines water depth, wave height, bottom conditions, surface currents, and wind/typhoon exposure to identify those regions posing the most difficult environmental problems, and hence requiring the highest level of technology for offshore operations.[48] Comparing the distribution of potential reserves (Appendix K) with the location of severe weather/high technology areas (Appendix N) leads to an unmistakable conclusion: the PRC's most promising offshore basins lie in those regions requiring the highest level of technology to exploit—the deeper reaches of the East China Sea. In most cases, these areas are beyond China's present drilling capabilities.

Development to date has, therefore, centered on the Po Hai and has included both foreign equipment purchases and domestically produced structures. The PRC now has a total of ten offshore platforms, not including their first deep water semisubmersible rig on order from the Fred Olson Group of Norway.[49] Of these, two are medium depth jack-up rigs purchased from Robin Loh Shipyard in Singapore, one or two are secondhand purchases from Japan, and the rest are Chinese-made. Table 4 gives a complete breakdown of platform type, characteristics, and location.[50] It illustrates that, with the exception of the two Singapore jack-ups, all Chinese activity has centered on the Po Hai, including at least one and possibly three production platforms.[51]

The Chinese also had three seismic survey crews in the gulf in 1975 and 1976, two equipped with native technology and the third with an imported navigation system and associated equipment.[52] In addition, there have been persistent rumors of PRC negotiations with Japan for construction of an underwater pipeline from gulf fields to an unspecified onshore location.

All these events suggest rather methodical self-development of the Po Hai by China. The PRC started serious exploration in the late 1960s, and despite one or two currently producing platforms, is essentially in the preproduction stage. Foreign purchases have filled technological gaps but have not represented the massive sorts of investments necessary to accelerate offshore development substantially. The emerging picture is one of the Chinese managing the entire enterprise alone and accepting the consequence of slower paced development. Although Western companies would welcome the opportunity to explore the Po Hai and adjacent areas, China as yet sees no reason to consider utilizing their services.

Table 4

China's Offshore Drilling Equipment

Rig Name	Type and Specifications	Location
1. Fuji	Jack-up. 300' water depth/ 15,000' drilling depth.	Po Hai
2. Kantan I	Catamaran. 250' water depth/ 800' drilling depth.	Po Hai/Yellow Sea
3. Kantan II	Catamaran. 250' water depth/ 800' drilling depth.	Po Hai/Yellow Sea
4. Pinhai I	Shallow water barge. 100' water depth.	Po Hai
5. Pinhai II	Shallow water barge. 100' water depth.	Po Hai
6. Pinhai III	Jack-up. 100' water depth.	Po Hai
7. Pohai I	Jack-up. 100' water depth/ 18,000' drilling depth.	Po Hai
8. Pohai II	Jack-up. 100' water depth/ 18,000' drilling depth.	Po Hai
9. Robray 2	Jack-up. 300' water depth/ 20,000' drilling depth.	South China Sea
10. Robray 3	Jack-up. 300' water depth/ 20,000' drilling depth.	South China Sea
Under Contract:		
1. Norway Rig	Semisubmersible. Capable of deep water (over 300') operations.	Unknown

Perhaps the best gauge of future PRC intentions offshore will be the rate at which semisubmersibles or similar deep-water rigs are acquired and manufactured. It is doubtful, however, that the recent purchase from Norway signals that a flood of semisubmersible buying is at hand. That purchase was concluded after two to three years of China's negotiating with other possible suppliers in Singapore and West Germany, and then only because it was a relatively new secondhand rig which could be obtained for a reasonable price.[53] Yet Chinese purchases of offshore well-logging equipment from United States suppliers and oil supply barges from Danish and Japanese firms, coupled with repeated references to simultaneous development of *all* oil resources at the April 1977 Ta-ch'ing Conference on

Industry, seem to indicate a cautious but steadily growing emphasis on offshore efforts. This strategy fits what many observers see as the most attractive pattern for China's waterborne activities: importation of very selected technology items for prototype use, plus a moderate but well-planned exchange of know-how and technical expertise with foreign companies.[54]

Despite this escalating interest in offshore areas, high costs, along with sophisticated technology levels and long lead times, promise to give the Chinese increasing difficulty in the years ahead. First, Western oilmen generally figure offshore development is three to five times as expensive as onshore production.[55] Not only has China yet to feel the full impact of these greater production costs, but it may also be faced, as discussed earlier, with geologic problems similar to those previously described for onshore fields. If much of the Po Hai represents an offshore extension of Ta-kang or Sheng-li, it is possible that the Chinese may again find an environment characterized by a highly fractured series of small, low-yield reservoirs whose cost of exploitation might escalate far beyond current expectations.

Second, whether China can locally produce or even absorb sophisticated foreign technology remains an open question. Observer reports indicate that existing Po Hai production platforms involve design technology roughly equal to U.S. platforms of the late 1940s to early 1950s. At several points along the gulf there is also evidence that the Chinese have had to reclaim land and set up conventional onshore operations due to apparent shortages of shallow-water equipment.[56] Perhaps these examples are not representative of China's technological potential,[57] but they do show that advanced capabilities cannot be assumed.

Finally, Western experts estimate that it requires five to seven years to reach substantial offshore production from a particular field:[58] three years from discovery of a commercial field to initial production and another two to four years before reaching maximum efficient output rates (MER). Oil industry planners tend to see comparable PRC development taking seven to ten years: four or five to construct platforms and associated piping and another three to five to achieve MER.

All of the above constraints could be mitigated by two factors. First, a Chinese decision to opt for service contracts or other joint arrangements with major oil companies could accelerate the entire development process. Yet domestic imperatives of self-reliance seem to argue against foreign contacts at present. Second, shallow water in the Po Hai should lessen all technical obstacles significantly. Depth and weather in the Po Hai, however, are not that different from conditions along the U.S. Gulf Coast, an area which generally has been subject to the same delays from cost and construction lead times despite benefitting from U.S. technology.

On balance, it appears that offshore resources will not make a meaningful contribution to Chinese crude production until the early to mid-1980s. Why, then, is China emphasizing this source so heavily? Lessened vulnerability to the USSR, political precedents for offshore claims, and proximity to potential markets are substantial but not compelling explanations; remaining onshore reserves and their locations are. If eastern fields will be unable to sustain Chinese production much past 1990, then a push either offshore or inland has to begin now. The huge infrastructure costs necessary to distribute interior production to markets and the proximity of any Dzungarian/Tarim basin pipeline to the Mongolian border would seem to indicate why the offshore alternative has been chosen.

4
People's Republic of China Domestic Variables

Before examining domestic considerations affecting PRC petroleum development, some brief comments on overall quality of information are necessary. In addition to previously described geologic uncertainties, descriptions of China's production potential reflect inherent observer biases. Japanese sources are frequently inclined to inflate their projections, based on a need to improve Japan's bargaining position with the Organization of Petroleum Exporting Countries (OPEC) and historical Japanese optimism in matters of trade with the PRC.[59] China is likewise prone to overstate out-year production estimates, thus improving its leverage in discouraging Japan's participation in Siberian development. Western oil companies, despite considerable reliance on pure economic and geologic asessments, either minimize or extol China's potential, depending on the likelihood of their eventually sharing the off-take.

Possible source bias is complicated by specific questions of completeness and veracity in China's case. Since 1960 the PRC has not published general economic statistics. The result is a series of oil industry production estimates based on aggregate percentage growth figures mentioned in the Chinese press and in the statements of Chinese leaders to visiting dignitaries; e.g., the level of production in year X was Y percent above the level in year Z. Although most scholars have concluded that the few PRC-released figures are accurate,[60] there is still persistent evidence of problems such as:

1. Deliberate falsification at lower levels of the Chinese bureaucracy to enhance performance statistics
2. Use of varying base periods to display percentage increases in the most favorable light
3. Complete withholding of data for bad years or poor performing sectors
4. Use of different definitions of product coverage and classification[61]

In short, information on China's petroleum industry is often incomplete and subject to a variety of political and systemic distortions. The only safe approach is to draw from the widest possible range of sources and to maximize cross-checking of all assessments.

Probably the most critical determinant of China's energy future is the domestic supply-and-demand picture. The CIA's 1975 study projected consumption growth rates between 1976 and 1980 by two methods[62] (see Table 5). The first uses of an energy/gross national product elasticity coefficient of 1.42 (1 percent GNP growth produces a 1.42 percent energy increase, the historical ratio for China since 1965) to calculate demand for different rates of economic growth. The alternate method extrapolates consumption trends for primary sectors and sums them to get total energy demand. The conclusion points to an annual increase of between 8.5 and 10 percent in *total* PRC energy demand from 1976 to 1980. This rate derives from rapidly rising consumption in the agricultural and industrial sectors and from moderately increasing energy requirements for transportation. From 1981 to 1985, due primarily to continued agricultural mechanization and industrial development, total demand growth looks to be in the higher 10 to 11.4 percent range.[63]

Table 5

Percentage Growth Rates

Elasticity Coefficient Method		Sector Projection Method
GNP Growth	Required Demand Growth	Demand Growth
4	5.7	8.4 (Low Case)
6	8.5	9.9 (Medium Case)
8	11.4	11.4 (High Case)

Of the principal reasons for rising consumption, agricultural mechanization is the most important. It was clearly spelled out as China's first economic priority by Hua Kuo-feng at a major PRC agricultural conference in late 1975, including references to a goal of "basic" farm mechanization in China by 1980.[64] Although such an ambitious objective appears beyond reach in that time frame, its implications in terms of energy, and particularly oil consumption, are enormous. Communes are already opting for machinery to smooth out peaks in labor demand and to eliminate a variety of low productivity tasks.[65] Use of diesel engines and electric motors is increasingly evident for numerous farm tasks, especially in irrigation and initial processing of agricultural commodities. As shown in Table 6, this trend is further accelerated by the PRC's practice of subsidizing diesel oil and electricity consumption for agriculture.

Finally, most observers believe that *the* major challenge for China's economy over the next ten to twenty years will be to achieve substantial and sustained increases in agricultural productivity.[66] Since Peking has already maximized the amount of suitable land under cultivation, it must obtain future growth almost entirely through programs of signifi-

Table 6

Relative Fuel Prices

Fuel	Purpose	Price ($)
Diesel	General	$0.20 per kg
Diesel	Agriculture	0.14 per kg
Gasoline	General	0.71 per liter
Electricity	Household	0.035 per kwh
Electricity	Industrial	0.03 per kwh
Electricity	Agriculture	$0.015 per kwh

Source: Amir V. Khan, "Agricultural Mechanization and Machinery Production in the People's Republic of China," *U.S.-China Business Review,* 3(6): 20, 23, November-December 1976. © The National Council for U.S.-China Trade, 1976.

cant technological change and intensive farm-related investment. The spread of double-cropping, the increased employment of chemical fertilizers, and greater use of irrigation equipment are but a few examples. Because additional yield increases will be accomplished primarily through greater energy inputs[67]—fuel, fertilizers, machinery, pesticides, and hybrid seeds—China's commitment to accelerate the mechanization process promises steadily rising *rates* of petroleum-related consumption for agriculture in the future.[68]

Growth in industry and transportation, while not as dramatic, will also contribute to increased demand. Industrial usage will come from expanding requirements for iron and steel production, chemical fertilizer plants, and electrical generation facilities, while the PRC's recent purchase of several petrochemical plants, all of which use petroleum products for fuel and feedstocks, will add an entirely new consuming source over the next several years. Transportation growth will be stimulated by progressive phase-in of diesel locomotives and by increasing use of trucks and automobiles.[69] For past and projected sector consumption rates, see Appendix O.

As total energy use has grown, the substitution of oil for coal, primarily in increments of new energy use, has increased. Table 7 illustrates the magnitude of this shift since 1957, correlates it with a similar shift in energy production to date, and, based on CIA figures, projects future source production percentages in 1980.[70]

Table 7

Energy Use and Production by Source

Use and Production	Past Percentiles				Projected Percentiles 1980		
	1957	1965	1970	1974	High	Medium	Low
Energy Use:							
Coal	94	85	76	69	--	--	---
Oil	4	8	14	21	--	--	---
Gas	1	7	10	10	--	--	---
Hydroelectric	1	Negl.	Negl.	Negl.	--	--	---
Energy Production:							
Coal	96	85	76	67	51	57	63
Oil	2	8	14	23	35	30	26
Gas	1	6	9	9	13	12	10
Hydroelectric	1	1	1	1	1	1	1

Although no future source usage rates are projected, past correlation between production and consumption percentages seems likely to continue. The pattern, then, is a combined coal and oil share of total consumption in the 85 to 93 percent range since 1965, with the coal-to-oil percentage ratio shifting from 85/8 in that year to 69/21 in 1974 to a probable 57/30 in 1980.[71] An even better example of substitution is in coal and oil portions of recent increases in total energy consumption. From 1958 to 1965, total energy usage increased by 1.16 million bpd (oil equivalent)—coal comprised 75 percent, while oil made up 12 percent. During 1971-1974, energy consumption rose by 1.36 million bpd—coal comprised 49 percent and oil made up 39 percent.[72] The conclusion is that shifts from coal to oil, plus rising total energy use, promise to keep Chinese petroleum consumption growing at 15 to 20 percent annually in the foreseeable future.[73] The key question is whether oil supply increases can continue to outpace this demand trend.

The size of future petroleum requirements, however, will be largely determined by China's success in increasing its coal supplies. Historic coal growth rates have been 6.5 to 7.0 percent annually, but recent increases have been considerably less: 6.3 percent in 1972, 5.9 percent in 1973, and only 3.2 percent in 1974.[74] Despite indications that this downward trend was reversed in 1975 and that subsequent troubles have been caused primarily by the T'ang-shan earthquake,[75] China's coal industry has a number of systemic difficulties which make maintenance of traditional growth rates unlikely.

First, coal has suffered from a relative lack of capital investment in both benificiation and extraction facilities for the last five to seven years.[76] Since beneficiation equipment is essential in cleaning the 20 to 30 percent of PRC coal used in the steel industry, its obsolescence slows down overall production and forces more lower grade coal to be supplied to non-steel industries.[77] Little capital investment for extraction means that the Chinese concentrate on raising the output of old mines through increasingly intensive applications of labor—a practice which often passes the point of diminishing returns and is a poor substitute for building new mines.

Another problem has been China's emphasis on localized diversification of the coal industry. A product of Cultural Revolution criticism, this strategy has produced a dramatic increase in the number of small coal mines serving only local needs.[78] Small mines have lightened the load on an overburdened transportation system, served as a catalyst for development of other local industries, and helped industry to better assist agriculture. But they have likewise failed to benefit from economies of scale and have produced lower quality coal, due primarily to primitive work conditions and little mechanization. These conditions, plus labor strife in mines and factories during 1974, were no doubt behind the convening of a National Coal Conference in 1975. The coal conference did not significantly alter China's general coal program, but it perhaps illustrated the scope of PRC difficulties by speaking of one or more Five Year Plans to overcome current supply short-falls.

A final problem is the geographical imbalance between China's iron ore and coking coal deposits. For example, northeast China contains only 1 percent of PRC high-grade coal but 13 percent of its iron ore; south central and southwest China hold 7 to 8 percent of the coking coal and 50 percent of the iron ore.[79] The result is that future coal and steel production growth will be increasingly tied to transportation improvements.

If coal's share of total energy production continues to shrink, rapid increases in the natural gas supply could help offset additional requirements for petroleum. As yet, the PRC uses little of this fuel outside of Szechwan Province, but greater use of gas associated with major oil fields is a relatively cheap way to fill growing urban demand. There is evidence that gas formerly flared at Ta-kang, P'an-shan, and Sheng-li is increasingly piped to such nearby cities as Peking, Tientsin, Shen-yang, and Tsingtao. However, the degree of substitution is limited. It depends on geographic proximity of field and market and is applicable mainly to new facilities constructed especially for gas. Gas may be regionally important but will not contribute much to the national supply picture without large investments in its distribution system and wholesale conversion of present coal-burning facilities. Accelerated demand, both in the aggregate and in gas-related industries such as fertilizers and petrochemicals, could force production greater than presently anticipated. But increased use of gas now flared, particularly from future offshore wells, would also carry the liability of lengthening construction lead time for the oil facility with which it was associated.

Table 8

Projected Levels of Total Energy Supply and Demand

Level	Projected Supply		Projected Demand	
	Oil Equivalent 1980 (Million bpd)	Annual Growth 1975-1980 (%)	Oil Equivalent 1980 (Million bpd)	Annual Growth 1975-1980 (%)
High	10.36	12.7	9.67	11.4
Medium	9.09	10.3	8.93	9.9
Low	8.03	8.0	8.21	8.4

Against the backdrop of sharply rising demand, greater oil-for-coal substitution, and lagging coal supply is a booming oil industry. Growing by 20 to 25 percent yearly until 1975, Chinese petroleum has so far offset these requirements and substantially expanded exports. Yet bottlenecks in related sectors, increasing capital investment requirements, and domestic political uncertainties remain to be dealt with. Sustaining 20 percent growth is further complicated by the normal byproducts of a longer production history: crude oil output increments will become more expensive as China exhausts cheaper sources; greater volumes of new production will be necessary to offset depletion from present wells; and percentage increases naturally tend to shrink as the statistical base grows. Indeed, the progressive slowing of 1975, 1976, and 1977 oil growth rates to 14, 13, and 9 percent,[80] respectively, indicates that such constraints may already be appearing.

Combining these factors into a total energy supply-and-demand matrix, CIA gets the alternatives for 1980 shown in Tables 8 and 9.[81] The projections show total energy supply slightly ahead of demand from 1975 to 1980 provided a medium supply level is maintained. Only in the unlikely high-supply growth case will China have a large exportable surplus.

It is more probable that a low to medium energy demand will be matched against a medium supply, implying a 6 to 6.5 percent annual GNP growth during the period and

Table 9

Net Total Energy Supply Levels in 1980
(Million bpd)*

Demand Level	High Supply	Medium Supply	Low Supply
High	0.69	-0.58	-1.64
Medium	1.43	0.16	-0.90
Low	2.15	0.88	-0.18

*Projected supply minus projected demand.

successful resolution of current coal supply problems. A GNP growth rate of more than 6 percent appears likely as lower rates would risk continuing the same economic problems which have plagued China for the past few years. The result will be 160,000 to 880,000 bpd of oil equivalent for export in 1980. When allowances are made for probable increases in the energy/GNP elasticity coefficient, the estimated exportable surplus will equal roughly 500,000 bpd.[82] Stronger domestic demand in succeeding years should place an absolute upper limit of 1.3 million bpd on exports in 1985, with a more likely level of *no exports* in that year unless the PRC (1) alters investment allocations sharply favoring energy production; (2) provides additional exports at the sacrifice of substantial economic growth; and/or (3) agrees to joint ventures with foreign firms for resource exploitation.[83]

Yet these and other supply-and-demand analyses have limitations. With little concrete data for the PRC, they rely heavily on projection of past growth rates leavened with a number of qualitative judgments about possible future developments. Unanticipated occurrences may never materialize or may cancel each other out, but before being discounted they should be examined in greater detail.

One such area involves potential supply bottlenecks in related industries which would affect petroleum. Capacity problems for coal have already been discussed, but lags in steel production could also have negative impacts. This now seems possible since insufficient recent plant investment has caused steel production to trail consumption since 1972.[84] As the PRC continues to mechanize agriculture and expand industry, it will require steadily growing quantities of steel in the years ahead and will probably have to increase both imports of technology and domestic resources allocated to production. Otherwise, China will face rising steel shortages in a number of key industries which would seriously hamper the production of exploration equipment, storage tanks, transportation facilities, and other petroleum industry essentials.

Another possible bottleneck involves China's transportation network. Since 1952 the PRC has obtained high rates of industrial growth by concentrating investment in producer goods industries such as energy and machine building.[85] Not only has previous development already realized most of the relatively inexpensive gains in those sectors (i.e., those involving low marginal capital output ratios), but it has also achieved such gains at the expense of transportation equipment and agricultural producer goods. Although the current mechanization push should remedy problems in agriculture, China's transport network remains hard pressed. With only 50,000 kilometers of railroads and 650,000 kilometers of highways in the early 1970s, it compares poorly with other major nations. The system represents a serious potential constraint at the margin, since it can both prevent higher growth rates in production and frustrate a more rational distribution of what is produced. Future improvements will involve fairly expensive investments which will not contribute directly to increases in output, but will instead simply facilitate such increases by eliminating serious distribution and delivery problems in the supply of inputs to producers. It is, in short, an area which will require a significant priority for out-year capital investment to avoid constraining several other sectors of the

Chinese economy.

The principal remaining bottleneck is the large investment in water storage and control facilities necessary, particularly on the North China Plain,[86] to obtain increased yields from greater applications of chemical fertilizer. The PRC has already undertaken the Yellow River Project to store water from the river, regulate its flow, and eliminate present silting problems. Even after its completion twenty years hence, however, this project will only add the equivalent of 7.5 percent to China's 1974 irrigated acreage. Considering the costs involved in irrigating just that portion of the North China Plain, and the probability that future yield increases must come from similar technological changes to existing farm methods, more extensive water control projects also appear to present a substantial hurdle to China's future economic development.

The above questions all revolve around the larger issue of general PRC economic priorities. These were perhaps best illustrated by Chou En-lai's January 1975 speech to the Fourth National People's Congress: the 1975–1980 Five Year Plan would mainly remove bottlenecks and give China's economy a base for takeoff; 1980 to 2000 would be a period of sustained high economic growth. Relative priorities seemed to favor agriculture, followed by light and heavy industry, with food, petroleum, and steel receiving most of the specific industry attention.

It seems that China plans on current resource allocation priorities being maintained until at least 1980. Yet, as shown above, emerging problems in iron, coal, and transportation may require significantly greater amounts of capital to avoid major bottlenecks.[87] Problems in steel are particularly troublesome, since emphasis on agricultural mechanization will place increased demands on that area for more trucks, tractors, and similar equipment. Moreover, any rise in consumer goods investment, especially if it releases pent-up demand for higher living standards by China's peasants, could easily skew current economic plans.

Even without these pressures, what investment will be required for China to sustain her recent oil growth rate of over 20 percent until 1980 and beyond? One source has estimated that it will take $20 billion per year to reach 3 to 5 million bpd by 1980 from onshore sources only; $7 to $13 billion per year to attain that production level by 1990.[88] Another has projected a $3.5 to $4.0 billion per year figure to reach 4 million bpd in 1980 and 6 million bpd in 1985.[89] For an economy whose 1976 GNP was approximately $250 billion, those kinds of investments for a single industry appear improbable.

Instead of citing overall estimates, a relative comparison of capital requirements for 20 percent annual oil growth might prove more useful.[90] This constant growth rate implies increments to output which increase in absolute terms and, assuming a constant capital/output ratio, growing levels of investment. Dr. K. C. Yeh has calculated 500 yuan (roughly $250) per ton of crude as a capital/output ratio for the 1950s. Since this included heavy initial infrastructure costs, a more conservative figure of 100 yuan per ton seems more accurate today—to avoid overstating the investment costs of increments to output.

With these investment/output ratios it is possible to compare the 1950s, a period during which China published economic statistics, with the 1970s and to determine any changes in capital investment percentages necessary for China to reach an output of 200 million tons (4 million bpd) in 1980. Using 100 yuan per ton, raising production from 99 million tons (2 million bpd) in 1976 to 200 million tons (4 million bpd) in 1980, would require a total investment of 10.1 billion yuan—roughly 5.3 times the 1.9 billion yuan invested in the petroleum industry from 1953 to 1957. By implication, the industry's investment share would equal 5 to 6 percent of the total invested in capital construction between 1976 and 1980 and would require almost *doubling* the average investment share of 3 percent that has recently (between 1971 and 1975) been devoted to oil production. Using identical methodology with slightly different base periods (1973 to 1977 and 1978 to 1982), the conclusion remains unchanged: oil's portion of China's total capital investment must roughly double over the next few years—from 3.9 to 7.4 percent—to maintain its recent 20 percent growth rate (see Appendixes P and Q).

Although these calculations are admittedly oversimplified and imprecise, they may at least define order of magnitude requirements. Whether investment increases will average 90 percent, as in each of the above cases, or only 50 percent is not that important. What is significant, however, is the apparent necessity for substantial increases in the normal resource share for petroleum during the very period when (1) coal, steel, and transportation are likely to demand larger investment percentages; and (2) agricultural development and associated mechanization plans seem to require similar relative increases. This situation seems especially acute since PRC access to world capital markets is somewhat restricted, requiring China to rely on internally generated investment resources to a greater extent than most other nations.

One caveat to a possible investment crunch is China's ability to mobilize vast amounts of cheap manpower. This reservoir of human capital has often helped the Chinese overcome past shortages in both industry and agriculture. Its potential for resolving future bottlenecks, particularly when China is confronted with seemingly absolute cost constraints, should not be underestimated.

Yet, while past infusions of unskilled or semiskilled personnel have succeeded in relatively low technology areas, their effectiveness in the oil industry is problematical. Built on the Soviet model, Chinese petroleum is a massive, horizontally and vertically integrated industry, controlled directly from Peking (see Appendix R). It is characterized by conservative innovation policies and relatively underdeveloped skill levels and possesses only a few design engineers of limited experience.[91] Its development to date suggests an attitude resistant to radical changes in equipment design or production flow patterns and an environment more conducive to complete plant imports than to large applications of indigenous labor.

Besides the doubtful efficacy of using a labor-intensive strategy for a technology-intensive organization, China faces major problems in raising its overall technical capacity. Traditional Maoist emphasis on "learning by doing" and mass education produces many low-to-mid-level technocrats but does not provide much advanced practical training. Particularly important is neglect of abstract, theoretical approaches to problem solving. Both in petroleum engineering research for improved production and in the geoscience work critical to exploration, lack of a theoretical framework inhibits drawing meaningful analytical comparisons.[92]

Post-Cultural Revolution policies have seemingly reinforced this trend toward applied learning. Educational reforms such as relaxation of university admission standards, greater political content in school curricula, and criticism of formalized grading systems seem to have lowered—or at least suspended—previous academic standards. Some of China's advanced technical institutes may be sheltered from these general reforms, but the evidence is not conclusive. This pattern may also change now that many of its leftist proponents have lost their former influence on educational policy. But such changes are not necessarily permanent and in any event will take several years to affect significant numbers of students.

Despite uncertainty over China's general technological capacity, the learning curve phenomenon is the PRC's most immediate technical obstacle. United States manufacturers report that the Chinese have read most published works in areas such as geophysical analysis but lack experience in practical application.[93] Even when importing the latest equipment, China may require several months or years to learn its proper operation. These shortcomings are often magnified by sheer technical complexity, so that despite possessing individual pieces of sophisticated foreign equipment, the Chinese lack the total systems technology to utilize those assets fully.

Although their recent acquisitions of Western equipment will be assimilated slowly, the Chinese are well aware of technology transfer and its potential benefits. Experience with Soviet aid during the 1950s showed them how the engineering techniques introduced through foreign equipment imports—and to a lesser extent the associated technical information exchange—played a key role in economic development. China's own literature suggests that the main problem during that period was not simple accumulation of domestic savings, but conversion of high savings rates, through the use of Soviet aid, into capital goods for heavy industry.[94] This experience demonstrates the importance of future petroleum equipment imports to the development of China's oil industry and suggests two threshold questions:

1. Can Chinese technicians completely assimilate such technology imports?
2. Will principles of self-reliance permit a sufficient influx of this technology to achieve significant petroleum production gains?

A related difficulty will be China's probable inability to use sophisticated material imports as development prototypes for domestic producers.[95] This will be occasioned not only by technological limitations but also by insufficient quantities of (1) high quality steel for the tubing, bits, and blowout preventers used in deep drilling; and (2) diodes, transistors, and other miniature electrical component parts necessary to repair imported geophysical equipment.

It is useful to leaven this pessimistic technology outlook, however, with a reminder about the limited forecasting ability of Western sources. When the Soviets withdrew their technicians from China in 1960, the PRC's chances of becoming a nuclear power appeared a decade or more away. But scarcely four years later the Chinese exploded their first strategic weapon, an event followed by rapid progress in all nuclear weapons areas. The anticipated technical hurdles for petroleum development, despite substantial contrary evidence, could prove equally ephemeral.

A final unknown in China's oil development equation is the PRC's domestic political future, with its immediate focus being the succession question. In other Communist states a durable Communist party has provided the necessary institutional framework for transition, but the Cultural Revolution and subsequent purge of heir-apparent Lin Piao have both weakened the Party apparatus and narrowed the circle of previously trusted leaders in China. Chou En-lai's death in early 1976 and Mao's subsequent passing will further complicate matters for China's new leaders—Chou's because he often served as a buffer between the chairman's thoughts and the realities of nation building; Mao's because his teachings have provided a well-understood sense of direction for all of Chinese society.

There are so many possibilities for change over the next few years, with competing sources of truth and no accepted interpreter of universal wisdom, that current forecasting is highly suspect. Although most observers consider it improbable, another Cultural Revolution is always possible. It would be consistent with Mao's own predilection for periodic upheavals as a means of renewing China's revolutionary fervor and of eliminating potential "revisionist" influences. While new Communist Party Chairman Hua Kuo-feng has purged his immediate rivals, the "Gang of Four," and has thus far successfully consolidated his power,[96] the sheer intensity of that struggle indicates the possibility for future discord. In addition, the generational split among some present and all future Chinese leaders complicates matters. Many of those currently in power are, like Mao, veterans of the Long March[97] and inheritors of the revolutionary legitimacy that it confers. The criterion used to select replacements for today's geriatric elite and their acceptance by a group which has yet to wield significant power are thus open questions.

In sum, the pride and sense of unity which have developed since 1949 should preclude any serious breakdown of authority, but successful transition into the post-Mao era should not be regarded as a foregone conclusion. Yet other factors exist which also seem to argue for stability. First, the absence of Mao himself as the unchallengeable instigator of revolutionary upheavals should probably lead to quieter times. Second, the last twenty-five years have demonstrated that radical policies are unsuccessful for more than short periods. Third, economic necessity and evidence of the damage wrought by past campaigns should push most groups in the direction of moderation. If any breakdown does occur, however, the resultant confusion and instability would likely have an adverse impact on economic efficiency in general and petroleum production in particular.

Notwithstanding the limited probability of another Cultural Revolution, two related factors may still affect future Chinese development. First, a series of minor political campaigns could adversely affect impact production in key industries. For example, the anti–Lin Piao/anti-Confucius campaign of 1974, factional strife over wage differentials in 1975, and particularly struggles surrounding the "Gang of Four" activities in 1976, all disrupted normal factory output levels. They seemed to show that local cadres panic if threatened with a mass political movement, quickly losing their authority and refusing to make the daily management decisions necessary to control grass roots production units. Petroleum has been well-protected to date, but continued expansion will invariably increase its susceptibility to such dislocations in the years ahead.

The second factor involves how various Chinese elites will interact over the long term. For ease of analysis, the normal

Table 10

Foreign, Domestic and Probable Oil Policies of Major PRC Elites

Policies	Militant Fundamentalists (LEFT)	Conservative Leftists
Domestic:	See necessity for constant class struggle and frequent mass mobilization campaigns; inclined toward total self-reliance in economic policy.	Seek ideological purity/conformity, but more within mechanism of Chinese Communist Party; favor periodic political mobilizations to rejuvenate but not dismantle Chinese political system.
Foreign:	1. Isolationist. 2. See USSR as greatest security/ideological threat. 3. More support for party-to-party relations abroad and increased support for National Liberation Fronts, even at expense of state-to-state relations.	1. Independence but not isolation. 2. Use of united front tactics for military but not economic purposes. 3. Avoid any form of economic dependence on other nations, even if PRC must forego much-needed technology. 4. About equal value assigned to international prestige and aiding NLFs.
Probable Oil:	1. Totally self-sufficient production with premium on meeting domestic demand. 2. Emphasis on economic autarky and tendency toward isolation that might curtail PRC export plans; would cut back/eliminate technology imports vital to rapid development of Chinese oil (especially offshore).	1. Would favor continued petroleum development, but mainly with indigenous skills and for internal needs. 2. Would likely put definite limits on Sino-US/Sino-Japanese trade levels and technology imports (which might translate into cutbacks from present levels). 3. Apt to support substantial oil exports only if they can be clearly linked to accomplishment of priority foreign policy goals (e.g., slowing Siberian development and consequent increases in Soviet strength by preventing Japan's participation; extracting political concessions from Japan on issues such as the pending Peace Treaty or Chinese offshore claims).
	Reform Bureaucrats (RIGHT)	Professional Military
Domestic:	Urge rapid modernization, stable and predictable rates for social change; respect for authority;	Emphasize stability, structure and order; advocate minimum interference by civilian sector and strong priority to heavy/defense indus-

Table 10 (Continued)

Policies	Reform Bureaucrats (RIGHT)	Professional Military
Domestic:	favor use of bureaucracy and no disruption of key strategic and economic sectors.	tries.
Foreign:	1. Activist approach favoring maximum use of united front tactics. 2. Stress strategic autonomy/increased national status thru economic development using imported technology. 3. Limited support for NLFs, but only as one of several elements in foreign policy. 4. Main emphasis on traditional balance-of-power strategy aimed at using Third World and its resources to restrict superpower flexibility.	1. Bias for retaining established relations over forging new ones. 2. Interested in agreements to increase trade/technology for upgrading defense posture. 3. Favor conflict avoidance unless China's vital national interests or territorial integrity threatened; has lukewarm support for NLFs in areas where they might provoke super power retaliation against China.
Probable Oil:	1. Would urge continued and possibly expanded oil exports to: a) Earn foreign exchange for increased technology/equipment imports. b) Increase regional political influence. c) Expand Sino-US/Sino-Japan ties (and possibly to enlarge PRC political leverage on specific bilateral issues). 2. Greater emphasis on imports to fill critical development gaps, but unlikely to abandon fundamental principles of self-reliance.	1. Difficult to assess; probably satisfied with present approach. 2. Might object if continued rapid oil growth diverted scarce resources from military preparedness. 3. Likely bias, similar to Conservative Leftists, in favor of demonstrable linkage to foreign policy objectives.

Left and Right divisions can be split into four basic groups: militant fundamentalists (Left), conservative leftists (Left), reform bureaucrats (Right), and the professional military (Right).[98] Table 10 on the preceding pages illustrates domestic, foreign, and likely oil-specific policies of each group. As is readily apparent, petroleum development options range from economic autarky, with little concern for oil exports, to accelerated production, with exports as a bargaining chip for a variety of political quids. The inevitable coalitions among these factions lead to further petroleum development paths. The point is not to forecast a particular outcome. It is rather to illustrate the wide range of potential oil production and export policies that the PRC might pursue, dependent on which groups or coalitions ultimately predominate now that Mao has passed from the scene. However likely, straight-line projections of oil production and export increases cannot simply be assumed but must be regarded as only one of several alternative policy paths between now and 1985.

The domestic variables discussed in this chapter all represent potential obstacles to China's oil development. Whether rising domestic demand, coal or steel supply problems, capital investment requirements, technological capability, or internal politics, they also seem destined to increase in magnitude in coming years. All but the ultimate succession question, however, are subject to amelioration by outside assistance, an alternative best viewed in relation to overall PRC trade policy.

5
People's Republic of China Trade

The central themes of China's trade policy are self-reliance and exchanges on the basis of equality and mutual benefit. The former seeks to preserve PRC independence by reducing reliance on outside assistance and by limiting foreign presence in China; the latter provides a means for the PRC to supplement its own resources while avoiding non-Chinese entanglements. In practical terms, China's imports are used to fill critical gaps in meeting the demands of agriculture and industry, while exports are used to pay for essential imports. This principle is perhaps best illustrated by actual trade patterns over the last several years: the volume of PRC trade has steadily grown, but its percentage of the GNP has remained almost unchanged.

Major trends since 1974 include the increased imports of entire plants and lesser scale machinery items initially, decreased and subsequently expanded exports of textiles and other normal export commodities, and greater use of deferred payments for import financing. During 1974 and 1975, worldwide recession precipitated a dramatic curtailment of traditional Chinese exports at the same time equipment import prices were rising. The PRC's response was to restrict imports by deferring purchases of less essential machinery and by reducing agricultural purchases over $800 million. China had previously tried to boost the price of textile exports, only to discover that recession-induced demand for such products was extremely elastic, leading to volume decreases which often exceeded the margin of price increases. Additional export constraints involved China's failure to research foreign markets closely and to tailor products to specific markets.

In spite of these difficulties, the PRC has steadily improved its overall trade position—from a deficit of $1 billion in 1974 to a projected surplus of $2 billion in 1977. The result was achieved by belt tightening in 1974 and 1975 and then by aggressively expanding exports of raw materials and finished products.

Although oil exports took up some of the 1974/1975 slack in PRC trade and will probably increase in future importance, China's recent rebound was accomplished against a backdrop of declining petroleum revenues. Oil earnings rose from $450 million in 1974 to a high of $860 million in 1975, but they then declined to $660 million in 1976 and are estimated to dip to $620 million in 1977.[99] Nevertheless, petrodollars have played a key role in financing over $2 billion of whole plant imports over the last three to four years, enabling the PRC to import needed steel, chemical fertilizers, and synthetic fiber plants without mortgaging its future to Western lenders. Greater use of oil earnings to finance technology imports seems probable, but other hard currency resources are also available. First, the demand for traditional PRC exports has risen as the West recovered from its recession. Next, the first of China's imported fertilizer plants began production in late 1976. For the remainder of the decade, these plants should help reduce China's need for chemical fertilizer and grain imports.

This sequence of events suggests that oil exports, while important, are far from an automatic, virtually unlimited source of foreign exchange. Extraction of sufficient resources to become a significant exporter would require large purchases of technology and equipment not currently available in China. The cost of these imports would theoretically be offset by future crude exports, but long amortization periods could lock Peking into exporting substantial quantities of petroleum to specific countries for extended periods, thereby constraining both its domestic and international freedom of action. Similarly, technology imports to stimulate oil output often involve time lags of several years from initial investment in exploration to eventual payoff in greater production. Such delays could confront the PRC with severe liquidity problems comparable to those recently experienced by Indonesia.[100] The upshot is that oil exports and the measures required to achieve them are no doubt a mixed blessing in Chinese eyes. Continued gradual increases seem probable, but any dramatic expansion could entail difficult political and economic choices for China's leadership.

If the incentives for expanded oil exports continue to increase, will China opt for foreign assistance in its petroleum development? In several ways the PRC has already agreed to limited cooperative arrangements with other countries. As Appendix S shows, the Chinese imported over $360 million in petroleum equipment alone between 1973 and 1977. This figure, in turn, represented only a small portion of total plant and equipment imports during that period. Peking has also permitted large numbers of foreign technicians to remain in China long enough to ensure proper plant installation and startup. A total of nearly 600 U.S. personnel alone, as detailed in Appendix T, have served in this function since 1973, and they were only part of the approximately 3,000 foreigners to have drawn similar assignments since that date.[101] In addition, about 20 PRC technicians spent four months during 1975 in Houston being trained in the use of geophysical equipment.[102] Other types of foreign training, plus frequent exchanges of visiting delegations in a wide variety of technical areas, likewise encourage more contact between China and its main trading partners. Finally, the Chinese have accepted use of medium-term commercial credits (up to five years for repayment) to finance major imports.[103] These and similar loosening trends have led some sources to predict that joint ventures, or at least equipment imports with eventual repayment in kind, are now inevitable.[104]

However, several considerations would seem to argue against such a radical departure from past Chinese practice. To begin with, such recent changes as deferred payments are mostly a resumption of emerging trends prior to the Cultural Revolution. No completely new policies have yet been adopted, and few seem likely as they could become the target of Leftist criticism. Additionally, any leaders espousing further liberalizations would run even greater risks in the post-Mao period, as no one of the chairman's stature could legitimize them. This policy may change if Hua Kuo-feng and the reform bureaucrats control China's future direction, but only after an extended transition period. Moreover, the oil industry in particular must contend with a Chinese legacy of exploitation. Whether the interest was Japanese, American, or British, the result was still a healthy off-take for the foreign user and little benefit for China.

Yet the basic resistance to outside aid comes from the nature of self-reliance itself.[105] It is first a means of avoiding dependence, prompted by a past history of colonial encroachments and, more recently, by overdependence on the USSR. It likewise involves Mao's effort to rebuild Chinese society in his own image, a political socialization process that strives to create a classless, unselfish "new Maoist man" dedicated to serving the State. This individual is totally confident of his ability to bend imported technology to his own ends and to master all development problems through his innate creativity and willpower. Self-reliance promotes this positive image while simultaneously limiting the corrupt influences that foreign presence might bring.

It is lastly a precautionary device intended to limit rising expectations. Since China's conceivable requirements for advanced technology are almost infinite—and certainly exceed its ability to earn foreign exchange—self-reliance exerts ideological pressure on industrial managers to rely on their own resources instead of clamoring for outside assistance. Such pressure is equally useful in reducing provincial, commune, or other local demands for resources from the central government. Furthermore, this practice of mobilizing local savings and limiting foreign influence has been relatively successful, enabling China to maintain a high degree of economic independence while holding imported technology to only 6 to 8 percent of total technological accretion.

The implications of self-reliance for China's petroleum development seem clear. Regardless of the type of transaction, China can be expected to refuse any form of direct foreign participation in its economy. In addition to specific policy statements to that effect, the PRC has repeatedly rebuffed joint-venture offers by various U.S. oil companies.[106] Other possible relaxations might include technical assistance agreements, production sharing, service contracts, or similar efforts which do not involve foreign equity interests. Increased borrowing and/or longer term credits might also be allowed, if only because no clear line exists between their adoption and current deferred payments practices.

Perhaps, given better relations with Japan and the United States and a Chinese-perceived need for accelerated development, technical assistance arrangements and greater use of credits would be possible. Indeed, recent PRC statements have mentioned using twelve-year deferred payments to finance technology imports, and, while again rejecting joint ventures, have referred to letting foreign companies "play a role" in the development of China's resources.[107] The Chinese might also make greater use of licensing arrangements for downstream processes and of specific contractual arrangements, especially for hiring specialized engineering advisors or leasing foreign equipment for exploration and initial resource development. But more fundamental concessions to self-reliance—a doctrine which is basic to Mao's teachings and which has served China well in the past—seem unlikely for the present. Pressure for such activities could mount after 1980, however, as the magnitude of China's petroleum production difficulties, particularly in the offshore area, becomes more apparent.[108]

6
Foreign Policy Questions

To this point, only China's ability to produce and export oil has been examined, yet the key policy question concerns the political significance of current and future Chinese oil production. China's oil diplomacy to date has been limited to a few specific actions:

1. The PRC has increased exports to Japan and looks forward to future expansion, partly to earn foreign exchange for Japanese equipment imports and partly to discourage Japan's participation in Soviet development of Siberia.
2. China has sold small amounts of crude, often at discounted prices, to Communist and non-Communist Asian states in an effort to preserve or enhance political influence.[109]
3. The PRC has used offshore drilling near its coast to demonstrate the seriousness of its territorial claims throughout the South and East China Seas, thus attempting to preserve its option for eventual development of those resources.
4. While endorsing OPEC actions, the Chinese have avoided direct association with the cartel and have

thereby benefitted from the OPEC's price policies without incurring any of its usual membership obligations.

Future directions of China's oil diplomacy will involve relations among all four major powers in Asia, with special emphasis on Soviet-Japanese and Sino-Japanese affairs, as well as general Chinese foreign policy objectives. Although this study's conclusions tend to minimize China's oil export potential and the consequent significance of any resource-based diplomacy, examination of the above areas should illustrate how the PRC will employ petroleum should a large exportable surplus develop.

Soviet-Japanese Relations

Post-war relations between the USSR and Japan have been characterized by slow, tortuous progress. While cold war alignments offer a partial explanation, widespread public dislike for the Soviets, Russian military strength, and a history of conflict between the two countries are probably more important. Factors affecting Soviet-Japanese affairs seem to center on three areas:

1. Japan's perception of USSR intentions and Soviet concerns about the future Japanese role in Asia
2. The dynamics of a long-standing dispute over Soviet possession of the Kurile Islands off northern Japan
3. The extent of Japanese cooperation in extracting oil and natural gas from Siberia

In the security area, each country has been extremely leery of the other's intentions. Japan has regarded the recent Soviet naval and strategic buildup with increasing concern and has been particularly troubled by the Soviets' advocacy of an Asian collective security proposal. Russian security concerns about Japan tend to be ambivalent. The Soviets would like to prevent closer relations between China and Japan and loosen Japanese ties with the United States, except that success might provoke the very type of independent Japanese military role they wish to avoid. So long as Japan is militarily weak, even a worsening of Soviet-Japanese relations would not pose a severe security dilemma for the USSR. But a Japan without U.S. defense guarantees or Chinese friendship would likely acquire its own military capability which, even if favorably disposed toward the Soviets, would be a source of potential problems.

The result is that the Soviets are in a distinctly unfavorable position, given the current configuration of forces in East Asia, to benefit from political maneuvering. They are locked into a pattern of conflict with the PRC while pursuing a negotiating strategy with Japan which has so far demanded much and yielded little. Yet a more liberalized negotiating posture is the only genuine Russian hope to compete for influence, since no other appeal will probably transcend Japan's community of interests with the United States or its cultural and political attraction to the PRC.

Energy and raw materials interests have dominated recent dealings between Japan and the USSR. Table 11 shows the scope of these interchanges which have involved Japanese commitments of approximately $1.7 billion to date.[110]

In addition to these undertakings, both governments have been discussing joint development of Siberian oil reserves since the late 1960s—a project whose erratic history best illustrates the problems inherent in Soviet-Japanese commercial relations. The initial Soviet proposal involved Japanese loans to extract oil from fields in the province of Tyumen in eastern Siberia and to construct a pipeline from that point to Nakhodka on the Pacific coast. Despite considerable interest, Japan proceeded cautiously because of questionable Russian reliability as a source of supply and in light of Chinese objections to resultant increases in Soviet

Table 11

Soviet-Japanese Projects Involving Energy and Raw Materials

Year	Project	Japanese Share	Remarks
1969	Forestry	$163 Million	-------
1972	Wood Chips	45 Million	-------
1974	Oil, Sakhalin Island	150 Million	Involves joint USSR-Japan offshore exploration off Sakhalin Island. Japan is to get lifting rights to 50% of any oil/gas discovered. Overall development costs should ultimately reach $1 billion.
1975	Coal, Chulman	540 Million	Coking coal for Japan.
1975*	Forestry	550 Million	-------
1975*	Ammonia Plants	245 Million	Total investment should eventually rise to $1.5 billion.
1975*	Gas, Yakutsk	$ 25 Million	Represents loan for drilling equipment and materials. Also involves $25 million parallel loan by Bank of America with USSR matching the other 50% of exploration costs. Full cost of project could eventually reach $7 billion, if sufficient gas reserves are discovered.

*Indicates amounts the Japanese are prepared to invest (if and when projects are implemented) rather than money already spent.

military capability. In early 1974 the Soviets abruptly raised Japan's cost share from $1.5 to $5 billion, reduced Japan's annual crude off-take from 800,000 to 500,000 bpd, and switched the requested delivery system from a large-diameter pipeline to a second trans-Siberian railway. Not only were the new terms much less attractive to Japan, but the railway also intensified PRC criticism. The result was eventual scrapping of the project, due primarily to its unattractive economics.

This stop-and-go process springs from the conflicting perceptions of each participant. The Soviets look to Japan as the cheapest available source of development technology and capital and as a stable market for Siberian resources. They also hope to increase their general political leverage on a resource-dependent neighbor while blocking further specific improvements in Sino-Japanese relations. Finally, Japan is the convenient catalyst for projects which will produce essential benefits for the USSR: increased energy supplies to offset rising domestic and East European demand, greater settlement of an underpopulated Siberia to strengthen Russian presence on China's northern border, and larger quantities of accessible fuel to increase the operational capability of Soviet ground and naval forces in the Far East. Although Japan's participation is probably not critical to final project completion, its assistance would substantially accelerate the development process and could obviate resource diversions from other areas of the Soviet economy.

The Japanese, meanwhile, must balance their need for supply source diversification, the possibility of energy cost savings, and the attractiveness of a potential Soviet market against PRC objections, questions of technical feasibility, and the likelihood of U.S. support to lessen the possibility of adverse changes in Soviet policy. In both cases, mutual distrust, coupled with external factors, seem to inhibit rapid progress.

Several conclusions emerge from this picture of Soviet-Japanese interaction. First, Chinese oil has not significantly affected Japan's involvement in Siberia. Despite the PRC's consistent attempts to link the two, its oil export prospects have simply not been great enough to generate much political leverage. China's objections have been considered by Japan, but progress has depended much more on direct Soviet-Japanese political relations and on the specific investment terms for each project.

Second, PRC oil is not really an alternative to Siberian energy: each is a possible way for Japan to diversify from Middle Eastern sources. Although future Soviet-Japanese-Chinese relations may limit the extent of Japan's participation in one or both sources, Japan will tend to view each on its individual merits. This seems evident even from the rather brief history of Soviet-Japanese energy projects. Japan is cautiously proceeding with Sakhalin offshore drilling,[111] the exploration phase of Yakutsk gas development, and the Chulman coal project, while discarding Tyumen oil mostly because of unfavorable development terms.

Third, sufficient roadblocks other than PRC concerns already exist to make Soviet-Japanese cooperation in Siberia a slow process yielding only limited benefits. A significant softening of the Russian negotiating posture could change this picture almost overnight, however, as the USSR has far greater resource potential than China. Such a change is unlikely in the near term, but could occur after 1980 if the Soviets feel greater need to discourage Japan's involvement with Chinese energy projects or to accelerate the exploitation of Siberia. The latter incentive would be especially compelling if recent CIA predictions about imminent declines in Soviet domestic production prove accurate.

Fourth, in the future Japan will experience increasing difficulty in balancing Soviet and Chinese interests. The PRC will continue to protest Soviet-Japanese energy projects and, due to its overall political importance to Japan, will reinforce that nation's natural inclination to proceed slowly in Siberia. Yet, if the Soviets offer more attractive business arrangements, Japan may be able to participate only at the risk of seriously alienating China. In the interim, a strong desire to avoid entanglement in Sino-Soviet polemics should prevent the Japanese leaning too far in either direction.

Sino-Japanese Relations

Relations between China and Japan derive from a legacy of shared interests and experiences. Cultural, linguistic, and racial similarities form traditional ties between each country, while Japanese guilt over wartime activities inspires a more recent sort of sympathy. Although this common heritage and history of mutual influence has caused past disagreements to acquire the ferocity of family feuds,[112] geographic proximity and economic complementarity make continued improvement in relations desirable, especially in terms of low transportation costs and exchanges of Chinese raw materials for Japanese equipment and technology.

After a postwar history of erratic and often bitter interactions, China and Japan normalized relations in late 1972. It was brought about largely by Chinese fear of Soviet intentions and was a natural follow-on to the Sino-American rapprochement earlier that same year. It also implied tacit Chinese acceptance of the U.S.-Japanese Security Treaty, at least until the Soviet threat subsided. The PRC apparently seeks to prevent a reduction of American forces in the Far East, since the USSR might fill any future void and since Japan, in the absence of credible U.S. security guarantees, might be increasingly inclined to expand its own military establishment. While Sino-Japanese relations are still on rather tenuous ground, successful consummation of accords on fisheries, trade, commercial air service, and navigation have fulfilled initial requirements of the 1972 normalization and seem to indicate a desire for continued progress on the part of both nations.

Dramatic increases in bilateral trade—from $1.1 billion in 1972 to $3.8 billion in 1975—illustrate the amount of recent progress. They are a function both of the improved political atmosphere, evidenced by the availability of Japan Export-Import Bank credits to China, and of basic economic factors. As mentioned above, Japan provides China with vital inputs for modernization while purchasing the raw materials and light industrial products which Japan's economy lacks or can no longer produce at low cost. China, in addition to needing finished steel, fertilizer, and similar products, also relies heavily on Japan for complete plants, the supplier credits to permit their purchase, and access to a wide range of development-accelerating technologies.

Yet current trade patterns are not without trouble spots.

Despite financing increased imports through oil exports, the PRC accumulated trade deficits with Japan of some $680 million in 1974 and nearly $1 billion in 1975. Concern over deficit size and lagging demand for some Chinese exports will help slow future trade growth, but more serious restrictions will probably come from the inherent limits on Sino-Japanese complementarity. China aspires to political autonomy through economic self-sufficiency and remains committed to radical egalitarian development policies in the Third World, whereas Japan relies on constantly expanding ing international economic activity and is dedicated to maintaining the stable political relationships necessary for increased trade and investment. In short, one nation sees independence and the other interdependence as the key to success in East Asia.

Within this context Chinese oil has emerged as the single most important trading commodity between the two countries. China's exports to Japan have risen from 20,000 bpd in 1973 to 160,000 bpd in 1975 and have been priced from a low of $3.75 per barrel prior to the Arab embargo to a high of $14.80 per barrel in early 1974.[113] (See Appendix U for terms of a typical Sino-Japanese petroleum supply contract.) In addition to increased oil imports, Japan has discussed possible cooperative arrangements with the PRC for liquified natural gas exports, undersea oil pipelines, and a host of offshore development projects. The Japanese have been particularly hopeful that China's increasing need for offshore technology will permit the compromises in self-reliance necessary for joint exploitation of the Po Hai and Yellow Sea areas.

Previous Japanese forecasts of 1980 oil imports totaling 1 million bpd, however, have proved optimistic. Negotiations on a long-term oil supply contract continued from 1974 through 1977, while recent exports declined to 122,000 bpd in 1976 and remained at that same approximate level in 1977. As of early 1978, the Chinese reportedly agreed to export

136,000 bpd to Japan in 1978 with the amount gradually rising to 300,000 bpd in 1982.[114] While specific amounts to be imported in intervening years will be negotiated separately, the 1982 figure apparently is a minimum level, thus giving Japan the option to increase the amount at a later time. In any event, this agreement likely represents only the most recent zigzag in the tortuous course of these bilateral oil negotiations, and, despite continued goodwill on both sides, economic realities appear to place an absolute upper limit of 500,000 to 600,000 bpd on Japan's imports of PRC crude through the early 1980s.[115]

Reasons for the import decline and the slow pace of negotiations are not hard to discern. First, Japanese refiners have just completed an investment program of over $1 billion to remove the sulfur from Middle East oil. Since as much as 40 percent of that capacity has not been fully utilized due to slack demand, the refiners would naturally prefer to import more high sulfur oil because (1) it is cheaper than low sulfur imports from China; and (2) it is the only way to amortize their desulfurization investment. Second, the Japanese do not want to abandon existing reliable supply sources, especially Indonesia, whose Minas crude competes with Chinese imports and which, as a future high growth area, offers Japan a potentially favored investment position it would find painful to sacrifice. Third, general lagging demand has biased Japan's petroleum industry against long-term supply commitments. Fourth, the high wax content of PRC crude requires specialized refinery cracking facilities to process. Japan's refiners lack this capability and are understandably reluctant to acquire it at the very time they are being financially squeezed by prior desulfurization outlays. Although Japan's Ministry of International Trade and Industry (MITI) has hinted at subsidizing part of the needed refinery conversion,[116] all the above factors make it improbable that Japan can greatly expand imports of PRC oil until at least 1980 or 1981.

Meanwhile, sluggish Japanese purchases might discredit Chinese factions which had originally urged oil exports as a near-term solution to foreign exchange and foreign policy leverage problems, with undetermined effects on future PRC decisions regarding capital investment in the petroleum industry.[117]

Japan's petroleum picture after 1980, however, looks better for the PRC. If overall demand steadily increases, more imports of Chinese oil will be urged by MITI to diversify sources of supply and by Japanese equipment manufacturers to facilitate export of their products to China. PRC oil would also be a more acceptable import than textiles, which compete with local Japanese industry, and would help improve general Sino-Japanese relations. The principal long-range variable thus appears to be Japan's post-1980 demand. The government's current five-year forecast, as detailed in Appendix V, foresees an annual petroleum demand growth of 4.9 percent between 1977 and 1981, while other longer range projections predict yearly increases of 6 to 7 percent between 1977 and 1985.[118] But crude and product demand may not grow as quickly as anticipated due to conservation and the substitution of coal and liquified natural gas (LNG) for oil, trends resulting in the displacement of products most easily made from Ta-ch'ing-type crudes and creating difficulties similar to those now being faced by Venezuela in exporting to the United States. In short, an open-ended Japanese market for PRC crude after 1980 cannot simply be assumed. It is instead dependent on a number of conditions:

1. Substantial amortization of present Japanese refiner investments in desulfurization equipment
2. Construction of new refinery facilities to "de-wax" Chinese crude
3. Sufficient Japanese demand, both overall and for heavier fuel oils

4. No adverse changes in Sino-Japanese political relations

In the event China does substantially increase its share of Japan's petroleum market, could Chinese political influence rise to the point of jeopardizing Japanese ties with other countries? The answer is probably no. China already has considerable influence with Japan because of geography, relative military strengths, and historical factors, but the risk of blackmail inherent in any Japanese resource dependency has been largely overshadowed by the 1973 Middle East war. Japan now finds it preferable to hostage at least some of its requirements to China, the USSR, and other suppliers rather than perpetuate its near-total reliance on Middle East sources. If no source is secure, the best alternative is to spread the risk. Japan, therefore, would likely encourage PRC imports up to a reasonable level (say, 10 to 15 percent of overall requirements instead of the current 2 to 3 percent share).

Yet the dynamics of Chinese-Japanese interaction are quite important to future expansion of China's petroleum industry. Japanese demand will not only encourage greater PRC exports, but Japanese technology will also help underpin Chinese production growth. Especially in offshore areas, the extent of Japan's technical, financial, and material assistance may greatly influence the pace of Chinese development. It follows that continued and possibly closer Sino-Japanese cooperation will be necessary in the years ahead if this mutually beneficial relationship is to endure.

How probable are continued harmonious relations during the next ten to fifteen years? On balance, the odds appear favorable. The value of Japan's trade is itself an indirect deterrent against PRC actions which might undermine the current rapprochement. In addition, both countries are strategically interdependent. Each shall continue to compete for influence in Asia with the instruments at its disposal, but neither should challenge vital interests of the other as both seek to suppress re-emergence of Japanese military capability. Moreover, China's desire to prevent the linkage of Japan's economic power with Soviet natural resources should prompt easier treatment of Japan on other questions. Finally, PRC fear of USSR intentions should continue to make China more tolerant of security arrangements between Japan and the United States.

There are, however, several possible trouble spots, any one of which could upset Chinese-Japanese affairs. First, Japan has substantial investment, trade, and defense interests in Taiwan and South Korea, areas where China has advocated or acquiesced to measures which undermine the status quo. In both these countries the presence of United States forces also involves U.S. and Japanese security ties. Second, Japan and China have made conflicting claims to offshore waters near the Senkaku Islands. If growing domestic demand were to absorb most of China's future oil production, its real need for promising offshore supplies near the Senkakus might conflict with Japan's requirement for a large indigenous oil source. This situation is further complicated by the predictions of Willums and others that these islands represent East Asia's single most promising offshore area. Even without an immediate need for Senkaku oil, the PRC might still feel compelled to press the issue as a precedent for other offshore territorial claims in the Far East.[119]

Third, China's growing nuclear arsenal risks stimulating Japan's own rearmament. Although Japan does not presently perceive this capability as threatening, future Chinese measures to enhance deterrence vis-à-vis the USSR will produce strong anxieties among conservative elements of Japan's political elite. If China took a militant position on Taiwan or other Asian areas where Japan's interests were involved, Japanese public opinion could shift toward acquisition of an independent defense ability. Fourth, Japan's continuing economic expansion throughout Southeast Asia leaves key Japanese interests potentially exposed to the interplay of local revolutionary forces. Should the PRC view this economic penetration as the precursor of political and military influence, it might be inclined to switch from merely "supporting" to actively "exporting" insurgent movements, particularly if Peking's moderates surrender some of their present leadership control.

This last issue has been persistently raised over the past few years by questions about the future direction of Chinese policy and the post-Vietnam credibility of U.S. alliances in Asia. The increasing military and economic capability of China to interfere in the affairs of others and the greater temptations to do so in the absence of a countervailing U.S. influence are both cause for Japanese concern. Yet such worries presently appear to be mitigated by several considerations. Chinese interest in state-to-state relations far overshadows any present support for local insurgent groups. This preference for dealing with established governments is increasing as a result of Chinese admission to the United Nations, the rapprochement with Japan and the United States, and other developments which enlarge China's stake in the existing international system. The PRC is also limited by intense political competition with the USSR; overt Chinese assistance to Asian revolutionaries could cause the target government to look for offsetting Soviet aid.[120] Furthermore, Japan's nonideological economic involvement in Asia tends to increase the acceptability of Japanese presence to Communist regimes—witness current dealings with Vietnam—while Japanese business is already expanding to Latin America, Canada, Australia, and similar locations. On balance, the chance of Sino-Japanese clashes in specific Southeast Asian countries appears remote, but this outcome depends largely on a shared perception—by China, Japan, third countries, and the revolutionaries themselves—that any insurgent success is relatively independent of Chinese policies.

On a more general plane, no outstanding Sino-Japanese issues need transcend the importance of gradual accommodation to both nations, but each will require considerable political skill in the years ahead to avoid mutual misperceptions and disputes which could lock the two into damaging regional competition in Asia. Neither country, however, has demonstrated such diplomatic agility in the past. Japan's history of calculated ambiguity and reactive foreign policy appears ill-equipped to handle an increasing number of either/or choices in the future—the possible choice between large-scale involvement in Siberia and Chinese goodwill being one example. China, meanwhile, is subject to abrupt policy changes based on the course of domestic policies and, due to the inward-looking emphasis of self-reliance, is not always an astute international observer. For instance, although China has not criticized Japanese militarism since 1972, this may represent more of a tactical shift than a real awareness of the dangers inherent in backing Japan into a corner.

Any of these potential developments could have serious consequences for Chinese petroleum development. Dependent on Japan for technology, capital, and an assured export market, China has to some extent mortgaged rapid production growth to continued cooperation with its Asian neighbor. Such sustained cooperation appears likely in the current Far East environment, but events in that region are seldom so predictable. Forecasts of harmonious Sino-Japanese relations automatically underpinning China's oil growth should be judged accordingly.

7
People's Republic of China Foreign Policy Goals

Before drawing conclusions about China's oil potential, it is useful to examine the possible role of expanded petroleum exports in Chinese foreign policy. Although major aspects of Soviet-Japanese and Sino-Japanese relations, plus recent PRC oil actions, have already been discussed, probable Chinese policies cannot be estimated without a larger look at China's general foreign affairs objectives. Significant Chinese influence based on oil seems unlikely, given this study's pessimistic view of export prospects, but even if future exports expand rapidly, gains should be limited as oil is only one of an increasing number of weapons in China's diplomatic arsenal. Its emergence may help influence, but should not predetermine, a particular course of action.

While necessarily oversimplified, past behavior suggests the following basic objectives for China's foreign policy:

1. Maintenance of adequate physical security by possessing sufficient conventional and nuclear forces to deter external attack
2. Reassertion of sovereignty over areas such as Taiwan considered to be integral parts of China
3. Establishment of an influence sphere around China

where neighboring countries at least do not threaten the PRC and, at best, adopt pro-Peking policies

4. Extension of Chinese influence throughout the Third World, both in the form of portraying China as a suitable socioeconomic development model and of supporting genuine revolutionary movements where feasible

In addition to these broader objectives, present PRC actions usually relate to a subset of attitudes toward specific countries:

1. Hostility toward the USSR, with the onus on Soviet initiatives to improve relations
2. Continued détente with the United States based on adherence to the 1972 Shanghai Communiqué
3. Continued friendship with Japan to benefit from technology transfers, to prevent Japanese acquisition of an independent defense capability, and to counter Soviet influence[121]

Before analyzing the effect of these goals on China's oil exports, two related considerations should be mentioned. First, to a larger extent than most countries, domestic politics control foreign affairs in China. Traditional Maoist emphasis on proper spiritual development and the current imperatives of nation building argue for the continued primacy of internal matters, a condition which simply magnifies the importance of China's leadership orientation in the post-succession period. Second, while the PRC now has a realistic, effective defense ability, recent emphasis on economic development has mandated military spending reductions of 20 to 30 percent from 1972 to 1974.

This trend seems likely to continue in light of the high absolute cost of new weapons systems, the high opportunity cost of diverting scarce trained manpower, and limited capital resources from the industrial and agricultural sectors. However, any dramatic changes in China's threat perception—whether the result of Soviet activity, American impotence, or other developments—would require substantial increases in the defense budget, possibly coming at the expense of oil-related investment.

Meanwhile, rivalry with the Soviets continues to be the centerpiece of Chinese foreign policy. China sees the USSR both as the prime external threat to its security and as the principal competitor for its international political position. Although the Sino-American rapprochement has diminished this threat, increasing Russian strategic and conventional forces, the Soviet collective security scheme for Asia, and recent USSR inroads on China's southern flank[122] all serve to reinforce previous historical and ideological enmity. China's main means of thwarting Soviet gains involve efforts to preserve competition between the United States and Russia and to preempt a strong Soviet-Japanese relationship. In the former case oil has a possible, though as yet undefined, role, while in the latter instance petroleum has already become an indirect, though as yet limited, instrument of Chinese influence. Past behavior suggests China might try to use increased oil exports in both circumstances to frustrate future Soviet aims, but the effectiveness of such a policy would depend on much greater PRC exports than are currently envisioned.

A similar pattern has emerged in regard to China's Third World interests. As a consistent champion of that group's policies, the PRC has openly endorsed OPEC's use of the oil weapon and has supported expansion of this technique to other raw materials.[123] In view of the traditional importance placed on rates of change in their assessment of world affairs, and since Third World fortunes are universally rising, the Chinese apparently visualize a future of progressively greater PRC power and progressively diminished Soviet-American preponderance, brought about in part by increased resource diplomacy. Yet, here again, modest export prospects and the availability of other foreign policy tools will probably limit China's potential influence.

In one sense, however, PRC influence has already experienced a quantum increase as a result of global concern over energy. With a primarily peasant economy and only limited means of projecting conventional military power, China has characteristically had to wield the promise of potential power and to play upon events beyond its own control to obtain its international goals. The 1973 oil embargo and subsequent worries about the security of existing energy supplies and the adequacy of new ones were ideally suited to this purpose. Peking could, and did, use its symbolic power as a future oil giant to promote a united front among Third World and Second World (Japan, Eastern and Western Europe) nations to lessen superpower domination and to thwart, or at least spotlight, Soviet political objectives. Petroleum exports simply offered tangible proof of the strong position from which the PRC spoke.

Overall, China's international goals offer mostly incentives for greater oil exports and suggest quite strongly how the PRC will attempt to use those exports:

- Shipments to Japan will probably continue to rise as (1) oil is one potential lever the Chinese possess for limiting Soviet influence in Japan and (2) supplying Japanese oil needs and increasing Chinese ability to purchase Japanese equipment are the best ways to strengthen overall cooperation between the two nations.
- Exports to Third World countries, particularly in Asia, will probably increase if China's production capacity outpaces future domestic and Japanese needs. Their selective use in these nations will be of frequent but seldom overriding importance, primarily due to modest PRC export prospects, quality problems associated with Chinese oil, and limited Third World refinery facilities equipped to process

such crudes.

- Token exports to the United States are a slim possibility. They would probably be more for symbolic purposes than for mutual economic benefit. An Alaskan-induced oil surplus on the U.S. West Coast and availability of other low-sulfur foreign crudes will likely render any PRC imports uneconomical, except at a substantially reduced price.

Whether or not exports sufficient to generate significant political leverage in the above ways actually materialize, the prospect itself should encourage China to keep expanding production levels beyond simple domestic requirements. To the extent such increases are achieved, they should increase Peking's stake in the international system, as well as the political and economic costs of any revolutionary foreign policy initiatives which might seriously upset the status quo.

8
Conclusions

Policy Implications

The future roles of PRC oil exports, whether large or small, seem clear. Although currently of secondary importance, a linkage between Chinese oil and Japanese participation in Siberian development does exist. In the near term it is motivated by China's fear of Soviet gains in Asia, of losing influence with Japan, and of potential increases in Russian military capability in the Far East. Lack of Japanese aid over the next 10 to 20 years, assuming continued absence of significant U.S. participation in Siberia, might contribute to a Soviet energy shortfall; Siberian oil will probably be needed to replace currently depleting reserves in more accessible parts of the USSR and to meet rising energy and raw materials requirements in both Russia and Eastern Europe in the 1980s. If Chinese oil can limit or preclude Japanese participation, exploitation of Siberian oil and gas would take longer, would require diversion of resources from other areas, and could force the Soviets to reduce energy exports to Warsaw Pact allies—all actions creating political and economic problems for the USSR.

Both short- and long-term results would benefit China.[124] First, Soviet economic dislocations would increase *relative*

Chinese economic and strategic strength, lessen the Russian military threat, and enhance China's political and ideological position in the Third World. Second, if Japan—or, for that matter, the United States—has only a small stake in Siberia, it will have less to lose in the event of a confrontation with the USSR and hence will be more likely to oppose future Soviet expansionism. The importance of this linkage should not be overemphasized, as the Soviets can theoretically turn to other sources of capital and technology. But in the absence of such changes, PRC oil remains a mechanism for discouraging Soviet-Japanese cooperation, as well as a possible means of diluting overall Soviet influence.

Another role of PRC oil is as a means of directly influencing Japan. Although Sino-Japanese relations have improved steadily since 1972, what if China suddenly reverted to its previous hostility toward Japanese alliance with the United States, only this time with the ability to make its will felt? Such an abrupt policy change is always possible, but it appears doubtful that oil alone will cause a significant increase in Chinese political leverage. China will remain the primary force in Japan's regional outlook, just as the United States will in Japanese global relations. These separate alignments will conflict on occasion, but Chinese oil should not greatly alter Japan's attitude toward either arrangement. The reasons for this sanguine outlook seem fairly clear. Even if Japan's intake of PRC crude reached 1 million bpd by 1980, this would represent only 15 to 20 percent of total Japanese petroleum imports[125]—a share which would likely be encouraged by MITI in order to diversify sources of supply.

The Japanese are also seeking out non-Arab energy sources in addition to the PRC. They are discussing oil imports with Venezuela and Mexico; they are engaged in ventures to develop Canadian tar sands and to prospect for Australian uranium; they are negotiating to participate in Vietnam's offshore development. In short, some Japanese diversification, whether to China or elsewhere, is beneficial as it decreases Japan's dependence on OPEC and its consequent incentive to break ranks with the United States and Western Europe in the event of another Arab embargo. Near-term potential for serious damage to U.S.-Japanese relations is probably greater from this source than from any increased dependence on Chinese oil, simply because Japanese-American interests presently come closer to converging on China than they do, or ever will, on the Middle East.

After 1980 the picture becomes more confusing, but, unless the PRC discovers wax-free oil or Japanese refiners convert their facilities, it looks as if competition from Minas crude and Japan's desire not to disrupt an Indonesian economic relationship of considerable importance will help prevent any excessive dependence on Chinese petroleum. In either case, China's limited export prospects should still serve to lessen PRC influence.

A final role of Chinese oil involves its possible use in accelerating linkages with the United States. If the development course described previously is correct, China will not be able to attain significant export levels without increased foreign assistance. Especially in offshore areas, the PRC will require sophisticated technology and equipment imports which could be best obtained from U.S. oil companies. Self-reliance would appear to rule out joint ventures, service contracts, or similar forms of cooperation, and, even if these arrangements were permitted, the prospect of PRC oil exports to American markets appears remote. Alaskan oil will create a crude surplus on the U.S. West Coast by 1978, making costly shipment to the U.S. East Coast the most probable alternative for Chinese oil. Even when compared to other low sulfur crudes (Alaska is high sulfur), China would have to compete in the West Coast market against well-established Indonesian oil, whose wax contents and transportation costs would undoubtedly be lower.

Despite these obstacles to direct resource transactions, the sale of oil-related equipment to China and U.S. company purchase of PRC crude for distribution within East Asia are distinct possibilities. Such dealings are also apt to become entangled with the entire question of PRC offshore claims and third party drilling concessions in disputed areas. On balance, anything more than token exports to the United States are unlikely, but the significance of Chinese oil for Sino-American relations should increase.

Development Implications

If these are the future functions of Chinese exports, what are China's chances of becoming a major exporter between now and 1990? At the beginning of this study, production estimates of 3.5 to 4.0 million bpd in 1980 and 8 million bpd in 1989/1990 were cited as the popular "consensus" figures of those who have thus far examined China's oil situation. These figures assumed a 20 percent annual growth rate from 1975 to 1980 and a 10 percent yearly increase from 1980 to 1990. Appendix W shows these and other predictions, as well as comparable Soviet production figures.

The Soviet figures are perhaps the best analog for long-term projections of Chinese oil production, as neither the Soviet Union nor China has permitted Western oil companies to help develop its resources. From 1957 to 1963, the Soviets experienced an average growth rate of 13.7 percent, and from 1964 to 1973, an average 7.5 percent increase.[126] When these growth rates are applied to production levels in analogous PRC years as in Appendix W, we see China producing 3 million bpd in 1980 and only 7 million bpd by 1990.

It may be argued that higher petroleum prices and greater technology levels today would raise China's growth rate in comparison to the earlier Soviet experience,[127] but several factors tend to offset any upward adjustments. First, observer reports indicate that most current Chinese equipment and technology date from the mid-to-late 1950s. Self-reliance, meanwhile, seems to put strict limits on the speed with which China can acquire later model machinery and processes. Second, higher oil prices are only a greater production incentive to the extent that they create foreign exchange earnings. However, quality problems and rising PRC domestic demand will probably restrict Chinese oil export earnings for the near future. Third, when the Soviets began expanding their crude oil output in 1954 and 1955, their supporting industries were much stronger than were China's in 1973 and 1974.[128] Fourth, post-1980 Chinese development must come increasingly from offshore or interior areas with all their associated cost ramifications. Russian production history includes almost no offshore production and only limited remote area development.

All this is not meant to endorse the Soviet oil model as a predictor of future Chinese experience. It is instead designed to illustrate the difficulty China will have in reaching the 4 and 8 million bpd plateaus in 1980 and 1990. This growth will require sustained, intensive effort and will remain dependent on the successful outcome of events in several possible problem areas, *any one of which* could derail attainment of the 1980 and 1990 bench marks. These problem areas can be categorized as follows:

Political

- Continued and even closer Sino-Japanese cooperation will be necessary throughout the entire fifteen-year period. Such a relationship seems probable under current conditions, but any number of future difficulties could arise. Specific trouble spots—where both China and Japan have divergent interests—include South Korea, Taiwan, and the Senkaku Islands. Other areas concern the exposure of Japanese economic activities elsewhere in Asia to Chinese-supported insurgent movements, the possible reversion of China to its pre-1972 hostility toward Japanese militarism, and the risk that a rising PRC nuclear capability might stimulate an independent Japanese direction in defense matters.

- The PRC must maintain its current "pro-development" domestic policies during the next fifteen years. For this to occur, the Chinese must forego the luxury of another Cultural Revolution, Great Leap Forward, or even a series of lesser campaigns for this entire period. China has not been characterized by such tranquility to date, however, and the succession question seems quite capable of provoking new upheavals. Without Chou En-lai to guide the reform bureaucrats and Mao to limit the scope of future mass campaigns, the possibility at least exists that China will again be subjected to economically disruptive internal turmoil.

Technical

- China will require increasing rates of investment from a relatively finite capital pool to sustain out-year production growth. Just to achieve 4 million bpd in 1980, the PRC will probably need to invest an *additional* $1.2 to $1.8 billion annually in its petroleum industry in the late 1970s.[129] This comes at a time when the push toward agricultural mechanization and obsolescence problems in coal, steel, and transportation seem to demand increased investment shares themselves. It is possible that greater capital demands can be met from rapid economic growth, but the risk that they may force an oil slowdown definitely exists. The problem will probably increase after 1980 as the threefold to fivefold cost factor for offshore development becomes apparent, and it will become especially severe if most Chinese production increments *above* 4 million bpd must come from fields other than those easily exploited in eastern China.

- Sufficient reserves must exist to support predicted output increases. Even assuming an optimistic reserve level of 20 billion barrels for east China basins, by 1978 to 1980

China will have to discover additional reserves at the rate of 3 to 5 billion barrels per year, just to sustain production over 4 million bpd. If recent predictions of PRC potential are correct, this discovery rate looks feasible, but lack of reliable geologic information and difficulties in mapping Chinese basins mean that the presence of adequate recoverable reserves cannot simply be assumed.

• China must successfully resolve a host of possible bottleneck problems. Coal supply must be increased at rates which will relieve the pressure on oil to fulfill ever greater portions of primary energy demand. Adequate quantities of specialized steel will be necessary, whether domestically produced or imported, to handle construction requirements for new pipelines, offshore platforms, refinery components, and the like. Requirements for additional transportation facilities, particularly pipelines for offshore oil and highways, railroads, and pipelines for interior fields will likewise strain Chinese capabilities.

• The Chinese labor force must acquire the necessary technological expertise to utilize recent innovative methods and equipment. China's education system will have to produce the large numbers of geologists, geophysicists, production engineers, and petroleum management specialists required to staff a major state-run oil industry. Their training will demand a technical proficiency level relatively unadulterated by post–Cultural Revolution reforms in education and will need to equip its graduates to overcome rapidly current learning gaps between China and the West.

These appear to be the main factors presently constraining Chinese petroleum development. The odds of any one preventing a production level of 8 million bpd in 1990—or an implied level of 5.5 to 6.0 million bpd in 1985—are probably less than fifty-fifty. However, taken together, they present quite a different picture. The key relationship is that all six possible problem areas must be positively resolved before the 8 million bpd figure can be reached.

Based on the foregoing analysis, the following individual probability values could be assigned each factor:

1.	Sino-Japanese cooperation	= .90	(90%)
2.	Pro-development policies	= .80	(80%)
3.	Sufficient capital investment	= .70	(70%)
4.	Adequate reserve base	= .90	(90%)
5.	Eliminate bottlenecks	= .80	(80%)
6.	Required technical expertise	= .70	(70%)

Each individual probability value simply indicates the author's estimate of the likelihood that successful event accomplishment—reaching 8 million bpd in 1990 or 5.5 to 6.0 million bpd in 1985—will *not* be prevented by that factor alone. Since each of the six factors is statistically independent of the others, the overall or joint probability is merely the product of all its individual probabilities.[130] The likelihood of China's achieving 8 million bpd in 1990, therefore, is approximately 25 percent:

$$P_{(Event)} = P_1 \times P_2 \times P_3 \times P_4 \times P_5 \times P_6$$
$$P_{(Event)} = 0.9 \times 0.8 \times 0.7 \times 0.9 \times 0.8 \times 0.7$$
$$P_{(Event)} = 25\%$$

Even if all six factors were given a 90 percent chance of success, the overall probability of event achievement would only be a little over 50 percent.

The above calculations are not intended to be a precise estimate of China's future production capacity: they are to illustrate the simultaneous nature of constraints on the Chinese oil industry and to show how that combined effect magnifies the difficulty of attaining dramatic output increases.

What is the significance of this conclusion? Is there any real difference between production levels of 5 or 6 million bpd in 1985, or of 7 or 8 million bpd in 1990? Will not China experience a substantial increase in economic and political power regardless of the precise output figure? Maybe, but differences in production must always be measured against accelerating domestic demand. If primary energy consumption continues to rise rapidly and petroleum is increasingly substituted for coal, China in the 1980s could have trouble maintaining a favorable energy balance. In a worst-case situation, the PRC might become a net energy importer by 1990. At a somewhat lower level, unrestrained demand might outstrip a supply of 5 million bpd in 1985, or 7 million bpd in 1990, forcing China to curtail exports or continue them only at the expense of significant economic growth. Indeed, most econometric studies of China's energy picture foresee exports approaching zero in the 1985-to-1990 period.[131]

These studies, in turn, are consistent with other observations which see nearer-term export levels governed primarily by PRC political decisions.[132] If China diverts oil from domestic markets to earn foreign exchange, for example, exports could rise to 500,000 bpd in 1980 and 1 million bpd in 1985. If the Chinese sell only that crude in excess of indigenous needs, exports would probably peak at around 400,000 bpd in the early 1980s.

But even if China generates an export surplus of 600,000 bpd in 1980 and 1.3 million bpd in 1985,[133] it is difficult to predict quantum increases in political influence. There will certainly be absolute gains, but whether relative power grows vis-à-vis Japan and the United States is an open question. China of the 1980s could well find itself more in need of development capital from those two nations than they are in need of Chinese oil. Differences in PRC output levels thus translate into varying assessments of Chinese ability to influence others in the next ten to fifteen years.

Analysis to date, comparing China to Saudi Arabia or other oil giants, has grossly exaggerated Chinese capabilities. It seems likely that even PRC capacity to use its oil weapon against Japan will be of only limited future significance. China will continue to wield considerable power over the next decade, but oil will comprise only one of the increasing number of diplomatic arrows in the Chinese quiver, complicating but not controlling the politics of East Asia.

Appendixes

APPENDIX A

MAJOR PRC OIL FIELDS

(In Thousands of bpd and Million Tons/Year)

Field (Geologic Basin)	Location	1974 Production	1975 Production	1976 Estim. Production	Remarks
1. Ta-ch'ing (Sung-liao)	North Central Manchuria.	680 (34)	780 (39)	850 (42.5)	Consists of over 4000 wells at average depth of 3000'. High residual & wax content.
2. Sheng-li (North China)	Near Yellow River entrance to Po Hai Gulf.	240 (12)	300 (15)	360 (18)	High residual & wax content. Significant quantities water.
3. Ta-kang (North China)	Adjacent to Tientsin.	60 (3)	80 (4)	100 (5)	High wax & residual content, altho less than Ta-ch'ing.
4. K'o-la-ma-i (Dzungarian)	In extreme NW China	30 (1.5)	30 (1.5)	30 (1.5)	
5. Yumen (Chiu-ch'uan)	No. Central China, northern Kansu Province.	20 (1)	20 (1)	20 (1)	
6. Leng-hu (Tsaidam)	West Central China	10 (0.5)	10 (0.5)	10 (0.5)	
7. Ch'ien-chiang	Hupeh Province, East Central China.	60 (3)	60 (3)	80 (4)	[Fields 7-10 are rough estimates. Production figures may be significantly higher or lower, altho these sources probably represent most of PRC's "unaccounted for" production. Fields are not shown on Appendix B.]
8. Fu-yu (Sung-liao)	North Central Manchuria.	60 (3)	60 (3)	70 (3.5)	
9. P'an-shan (North China)	Liao-ning Province, NE of Peking near Liao River.	60 (3)	60 (3)	80 (4)	
10. I-tu (North China)	Just south of Sheng-li field.	20 (1)	20 (1)	20 (1)	
11. Shale Oil	Various	60 (3)	60 (3)	60 (3)	
	TOTALS:	1,300 (65)	1,480 (74)	1,680 (84)	

Sources: JEC, Williams, "The Chinese Petroleum Industry," *op.cit.*, pp.233,250-60; U.S.Dept. of Interior, Bureau of Mines, K.P.Wang, *People's Republic of China: A New Industrial Power With a Strong Mineral Base* (Washington:GPO, 1975), pp.39-45; *Far Eastern Economic Review*, 90(45):34, Nov. 7, 1975; *Washington Post*, Dec. 12, 1975; Chu-yuan Cheng, *China's Petroleum Industry: Output Growth and Export Potential* (N.Y.: Praeger Publishers, 1976), pp.32-79; B.A.Williams, "The Petroleum Industry in China: A Note," Mar. 1976, pp.6-10, Table 2 [unpublished]; U.S.CIA, *China: Oil Production Prospects*, ER 77-10030U, June 1977, pp.9-10; A.A. Meyerhoff and Jan-Olaf Willums, "Petroleum Geology and Industry of the People's Republic of China," UN, Economic and Social Commission for Asia and Pacific, Committee for Coordination of Joint Prospecting for Mineral Resources in Asian Offshore Areas (CCOP), *Technical Bulletin*, Vol. 10, December 1976.

APPENDIX B

MAP OF PRC OIL RESOURCES AND REFINERIES

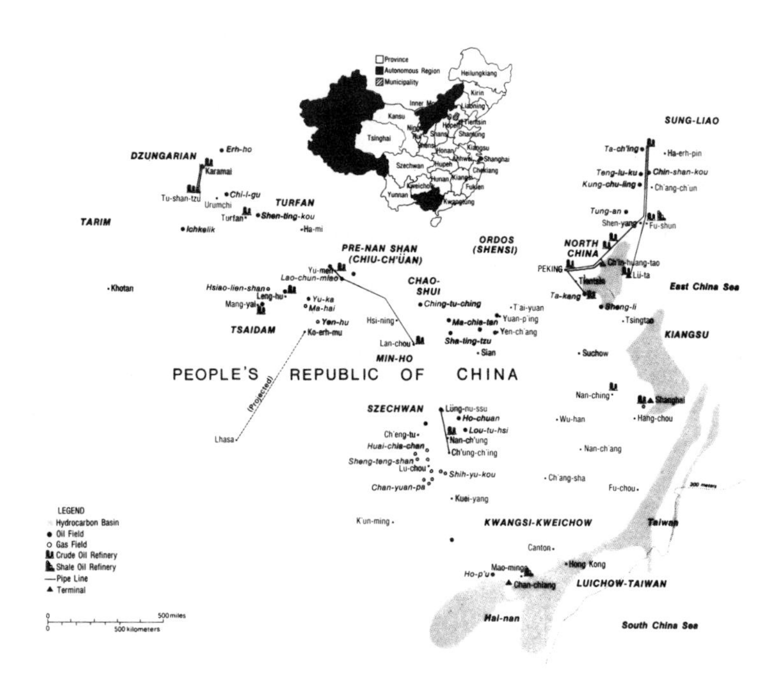

Source: *World Oil*, October 1975, p. 145.

APPENDIX C

OVERALL ESTIMATES FOR ONSHORE AND OFFSHORE RESERVES

Oil Basins	Meyerhoff (1970)[1]	Meyerhoff and Willums (1976)[3]	Major No. 1[4]	Major No. 2[5]	Ho K'o-jen[6]	Private Study[7]
A. Interior Basins:						
1) Dzungarian	880 MMB	4.88 BB	10 BB	---	0.89 BB	---
2) Tarim	50 "	5.05	20	---	--	---
3) Tsaidam	1,760 "	5.32	10	---	8.36	---
4) Pre-Nan Shan/Chao-shui/Min-ho	633 "	1.70	2	---	0.52	---
5) Ordos	29 "	0.24	5	---	0.37	---
6) Szechwan	1,116 "	3.06	8	---	0.59	---
7) Turfan	50 "	0.50	--	---	--	
Total Interior:	4.51 BB	20.75 BB	55 BB	20 BB	10.73 BB	22.6/41.2 BB
B. Eastern China Basins:						
1) Sung-Liao	744 MMB	9.00 BB	10 BB	---	0.89	---
2) North China (Onshore)	385 "	6.76	10	---	0.30	---
3) Kwangsi/Kweichow	---	2.00	0.5	---	--	
Total Eastern China:	1.16 BB	17.76 BB	20.5 BB	10 BB	1.19 BB	18.8/27.7 BB
Total Onshore China:	5.67 BB	38.51 BB	75.5 BB	30 BB	11.92 BB	41.4/68.9 BB
C. Offshore Areas:						
1) Po Hai (North China Basin)	--	5.50 BB	10 BB	6 BB	--	---
2) Yellow Sea (Kiangsu Basin)	--	4.50	10	4	--	---
3) East China Sea (Kiangsu Basin)	--	20.00	UNK	UNK	--	---
Total Offshore:	--	30 BB	20+ BB	10 BB	--	---
Total Onshore + Offshore:	19.6 BB[2]	68.50 BB	95+ BB	40 BB	--	---
East China % Total Onshore Reserves:	20%	42%	27%	33%	10%	45/40%
Interior % Total Onshore Reserves:	80%	58%	73%	67%	90%	55/60%
East China % Total Chinese Reserves:	--	22%	21%	25%	--	---
Interior % Total Chinese Reserves:	--	30%	58%	50%	--	---
Offshore % Total Chinese Reserves:	--	48%	21%	25%	--	---

1. Meyerhoff, *AAPG Bulletin*, Aug. 1970, *op. cit.*, pp. 1570-73.
2. *Ibid.* Meyerhoff arrives at total proved+probable+potential reserves figure of 19.6 BB, by:
   a) Adding to 5.67 BB (probable ultimate recovery from known fields) another 1.33 BB; which represents oil-bearing zones known to be present in several fields, but not yet produced, for total of 7.0 BB estimated proved + probable reserves in known fields.
   b) He then calculated another 12.6 BB of potential reserves in undeveloped structures.
   c) a + b = 19.6 BB total, which Meyerhoff emphasizes is a minimum or "conservative" estimate.
3. Meyerhoff & Willums, *CCOP Tech. Bull.*, *op. cit.*, pp. 201-02. Represents estimated ultimate oil recovery from basins (proved+probable+potential reserves). Absolute reserve figures reduced by oil already produced from PRC basins (roughly 2.8 BB from Eastern fields; 0.3 BB from Interior) before calculating current reserve ratios for different geographic areas.
4. Represents an "optimistic" or ceiling estimate of a geologist at one major oil company.
5. Represents overall assessment of another major oil company.
6. Ho K'o-jen, "Developments in Red China's Petroleum Industry," *Fei-ch'ing Yueh-pao* (Taipei, Mar. 1, 1968), pp. 10-11, trans. Joint Publications Research Service, U.S. Dept. of Commerce. Includes Class A and B onshore reserves only (roughly equivalent to proved and probable reserves).
7. CIA, *China's Oil Prospects*, *op. cit.*, pp. 5-7. The study was an analysis by a group of petroleum consultants. Figures on left indicate proved + probable reserves; those on right indicate proved + probable + potential reserves.

APPENDIX D

ESTIMATE OF SIZE OF NEW OIL RESERVE FINDS NECESSARY TO MAINTAIN R/Ps OF 15 and 30 (in Billion Barrels)[1]

(A) Year and Production Growth Rate	(B)[2] Annual Production	(C) Cumulative Production	(D) Reserves at Year-End	(E) Reserves Necessary For An R/P = 15	(F)[4] Annual Finds Needed to Maintain An R/P = 15	(G) Reserves Necessary to Maintain An R/P = 30	(H)[5] Annual Finds Needed to Maintain An R/P = 30
1974 (Actual)	0.43	0.43[3]	20	6.4	0	12.9	0
1975 (Actual)	0.53	0.96	19.47	8.0	0	15.9	0
1976 (15%)	0.64	1.60	18.83	9.6	0	19.2	0.4
1977 (15%)	0.74	2.34	18.09	11.1	0	22.2	3.7
1978 (15%)	0.85	3.19	17.24	12.8	0	25.5	4.2
1979 (15%)	0.99	4.18	16.25	14.9	0	29.7	5.1
1980 (15%)	1.17	5.35	15.08	17.6	2.5	35.1	6.7
1981 (10%)	1.25	6.60	13.83	18.8	2.5	37.5	3.6
1982 (10%)	1.37	7.97	12.46	20.6	3.1	41.1	4.9
1983 (10%)	1.50	9.47	10.96	22.5	3.3	45.0	5.4

1. This chart uses only the *reserve base* in *eastern* PRC fields to measure R/P requirements, since that region will likely remain China's predominant producing area until at least the early 1980s. Again, the "optimistic" 20 BB total reserve figure for eastern fields is used to provide a liberal margin for error and still illustrate the magnitude of additional reserves China will require, either from offshore or interior basins, if it is to sustain annual increases in production of 10 to 20%. For similar methodology using a conservative initial reserve base for all of China, see Tatsu Kambara, "The Petroleum Industry in China," *China Quarterly*, October-December, 1974, pp. 717-718.

2. Figures exclude the roughly 10% of Chinese production that already comes from interior fields, and assume that the percentage will remain constant through 1983.

3. No effort was made to compute cumulative production's depletion of reserves prior to 1974, as most reserve estimates have been made since that erosion took place.

4. (F) = (E) - (D) - cumulative of (F) to previous year.

5. (H) = (G) - (D) - cumulative of (H) to previous year.

APPENDIX E

PRC REFINERIES[1]

Location	No. Refineries at Each Location	1975 Capacity (Thousand bpd)
1. Sheng-li	3	144
2. Fu-shun	2	130
3. Ta-ch'ing	3	118
4. Ta-kang	3	84
5. Lan-chou		80
6. Dairen		80
7. Shanghai		80
8. Peking		80
9. Chin-hsi		70
10. Nan-ching		60
11. Lin-hsiang		50
12. An-shan		50
13. Mao-ming		50
14. Fu-yu		40
15. Ching-men		40
16. Wu-han		34
17. Pao-ting		30
18. Tu-shan-tzu		30
19. Yu-men		20
20. Hang-chou		18
21. Yang-liu-ch'ing		16
22. Nan-ch'ung		10
23. Leng-hu		6
24. Others		10
	Total:	1,330

[1]Chart represents a composite from the following sources: JEC, Williams, "The Chinese Petroleum Industry," *op. cit.*, pp. 244-245, 263; Wang, *The PRC: A New Industrial Power*, *op. cit.*, p. 35; *Petroleum Times*, July 11, 1975, pp. 26, 35; Williams, "A Note," *op. cit.*, p. 16; CIA, *China Oil Prospects*, pp. 11-21; *China's Petroleum Industry*, *op. cit.*, pp. 37-39.

APPENDIX F

CHINESE REFINING CAPACITY AND CAPACITY UTILIZATION

Year	National Crude Output	Crude Net Imports	Crude Consumed at Power Plants	Crude Requiring Refining[1]	Average Refining Capacity[2]	Utilization of Capacity
			(Thousand bpd)			
1965	220	2	---	210	272	78%
1966	282	2	---	270	356	76
1967	278	2	---	266	400	66
1968	304	2	---	290	456	64
1969	408	2	- 20	370	520	71
1970	564	10	- 20	528	630	84
1971	734	4	- 40	664	760	87
1972	862	2	- 60	764	912	84
1973	1,096	- 4	- 80	960	1,088	88
1974	1,316	- 70	-100	1,086	1,156	93
1975	1,486	-154	-120	1,154	1,330	87%

[1]After a 5% transport and refining loss.

[2]The arithmetic average of year-end figures.

Source: CIA, *China Oil Prospects*, p. 26.

APPENDIX G

MAJOR FIELD CRUDE OIL CHARACTERISTICS AND AVERAGE PRODUCT YIELDS

Crude Oil Characteristics	Ta-ch'ing	Sheng-li	Ta-kang[1]
Gravity, °API	32.8	20 - 24.6	---
Wax, Hexane (% wt.)	22.4	15.3	10.0
Sulfur (% wt.)	0.11	0.98	0.2
Water Content (% vol.)	0.5	0.7	---
Pour Point (° C)	(+) 35	(+)27.5	---
Average Product Yields (% volume)			
Liquid Propane Gas	0.2	---	--
Light Naptha	5.4	---	--
Naptha	3.7	7.3	--
Kerosene	4.7	5.0	--
Gas Oil	6.9	9.6	--
Heavy Gas Oil	7.0	---	--
Residue	68.0	77.1	--
Water	2.6	---	--
Loss	1.5	1.0	--

[1]Complete data are not available on Ta-kang crude. Its characteristics, however, are reported to be similar to those of Sheng-li crude.

Sources: CIA, *China Oil Prospects*, *op. cit.*, pp. 5,11,13; Meyerhoff and Willums, *CCOP Tech. Bull.*, *op. cit.*, p. 190; *China's Petroleum Industry*, *op. cit.*, p. 53. The latter reference reports its data are from a 1974 survey of Japanese industry sources, based on spot analysis of single samples.

APPENDIX H

WORLD CRUDE OIL RESIDUAL YIELDS[1]

Crude	Origin	Gravity °API	End Temperature (°C)	Residuum Percentage
Arabian Light	Saudi Arabia	33.4	343	46.1
Iranian Light	Iran	33.5	343	45.4
Dairus	Iran	33.9	327	41.03
Basrah	Iraq	33.9	343	44.54
Kuwait Crude	Kuwait	31.2	360	47.53
Minas Crude	Indonesia	35.2	343	56.5
Sassan	Iran	33.9	371	39.7
Gulf of Suez Blend	Egypt	31.5	366	47.0
Reforma (Cactus Reforma Isthmus)	Mexico	33.0	343	45.3
Trinidad Blend	Trinidad-Tobago	33.6	343	31.2

[1]Source: CIA, *China Oil Prospects*, *op. cit.*, p. 8. The initial process of refining involves heating crude oil in a distiller which bleeds off constituents of the crude according to the temperatures at which they vaporize. That portion of the crude left unvaporized, after reaching the highest practical distiller temperature, is the residuum. As shown in Appendix G and note 40, Ta-ch'ing and Sheng-li crudes have residuums of 68 - 70 percent and 77 - 84 percent, respectively.

APPENDIX I

CHINESE CONTINENTAL SHELF

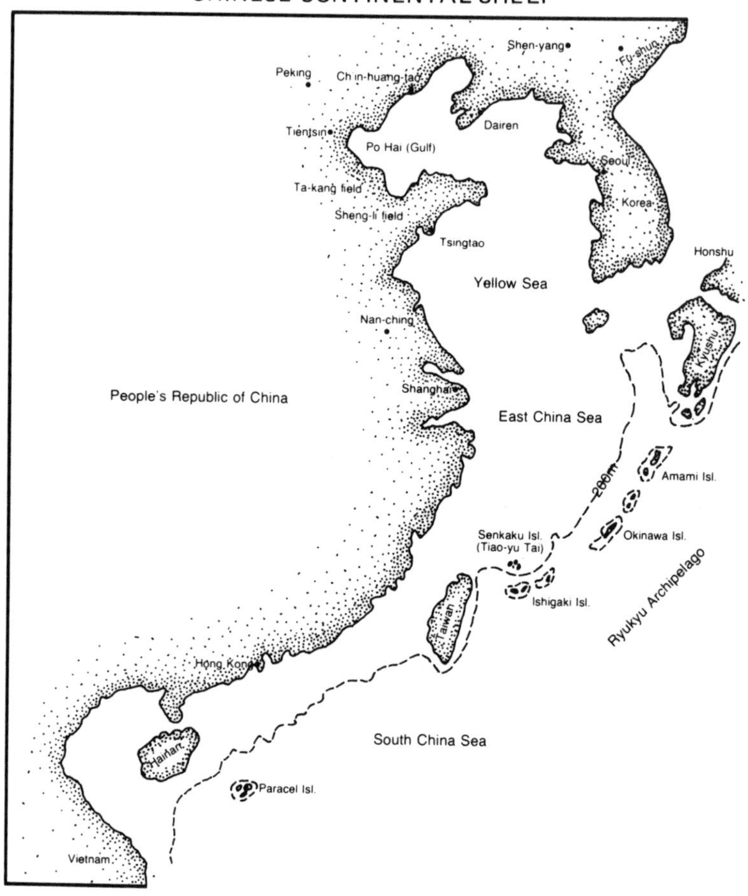

Source: Jan-Olaf Willums, "*Prospects for Offshore Oil and Gas Developments in the People's Republic of China*" paper presented to the Sixth Annual Offshore Technology Conference, May 1974.

APPENDIX J

ESTIMATED HYDROCARBON POTENTIAL OF THE CHINESE CONTINENTAL SHELF

Area Name	Technology Class	Total Surface Area	Total Volume	Total Recoverable HC Potential			Total In Place HC Potential		
		km^2	km^3	Pess. 10^9bbl	Middle 10^9bbl	Opt. 10^9bbl	Pess. 10^9bbl	Middle 10^9bbl	Opt. 10^9bbl
1. Gulf of Po Hai and Korean Bay	I	161'300	383'000	1.3	3.5	13.1	5.2	14.0	52.9
2. Yellow Sea	II	162'100	494'000	1.6	4.2	15.8	6.4	16.8	63.2
3. East China Sea Shallow Section	II	105'500	135'000	0.1	2.1	60.0	0.4	8.5	240.0
4. East China Sea Deep Section	III	216'100	743'000	3.7	10.4	175.0	14.8	41.6	700.0
5. Formosa Strait and Taiwan Area	II	112'900	330'000	0.8	3.4	7.6	3.2	13.5	30.4
6. South China Sea Shallow Section	II	58'900	150'000	0.2	1.0	2.3	0.8	3.9	9.2
7. South China Sea Deep Section	III	157'000	470'000	0.7	3.0	6.7	2.8	12.2	26.8
8. Gulf of Tonkin	I	78'000	191'000	0.3	1.4	3.1	1.2	5.4	12.4
Total:		1'051'800	2'896'000	8.7	29.0	283.6	34.8	115.9	1134.9

APPENDIX K

CHINESE OFFSHORE HYDROCARBON POTENTIAL

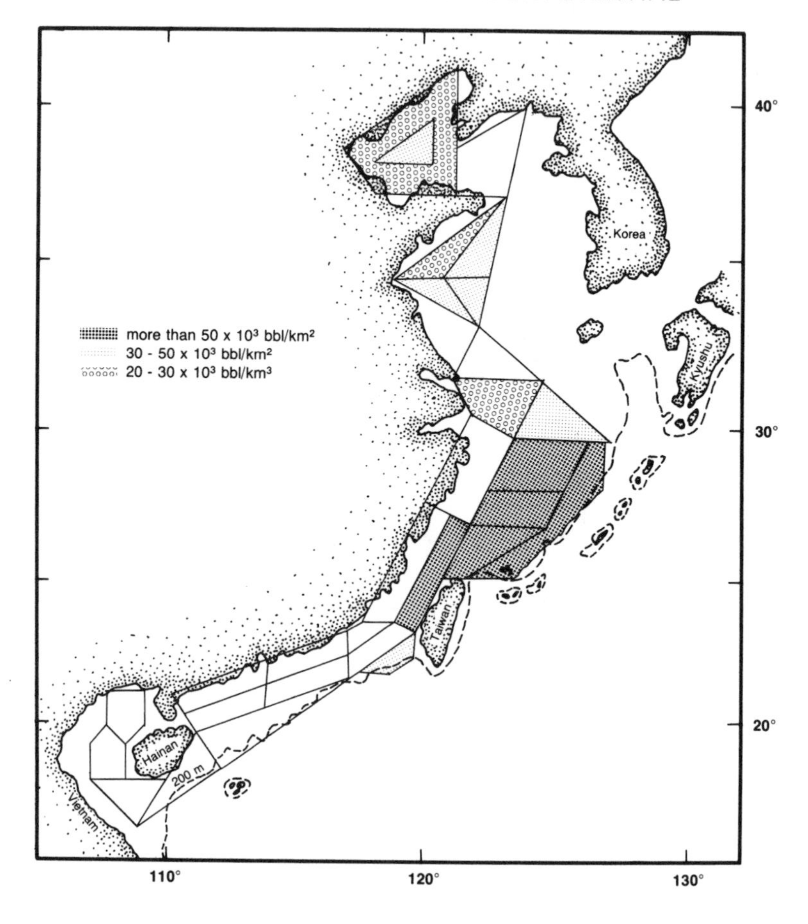

APPENDIX L

SAMPLE OFFSHORE CLAIMS

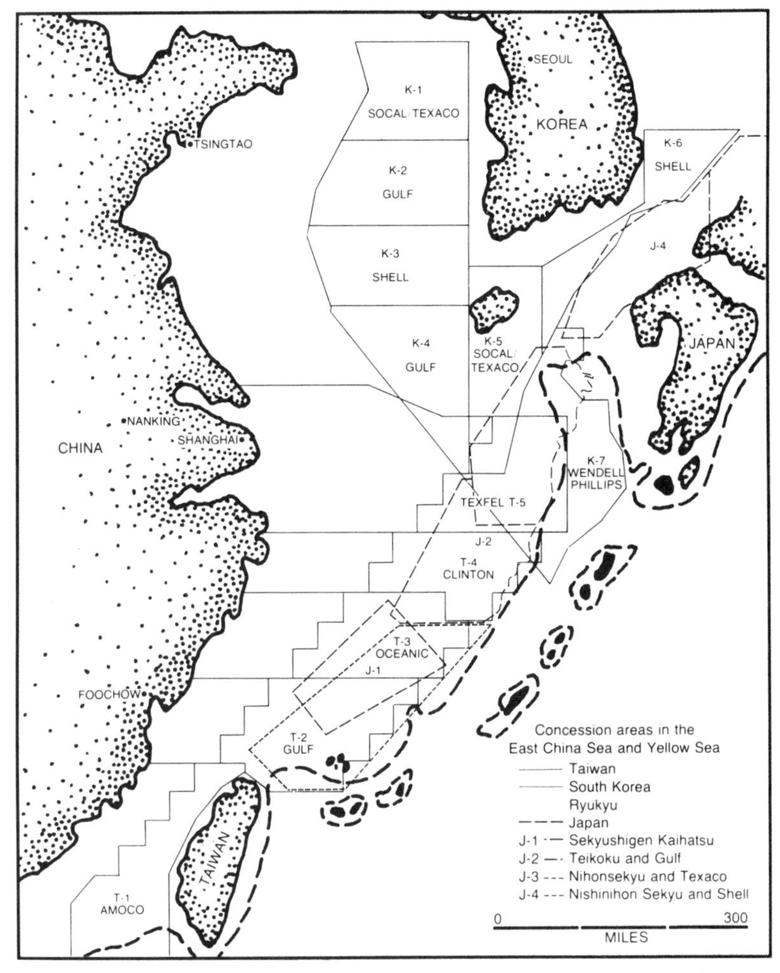

Based on U.S. State Department Map No. 261 7-71 State (RGE), with additions.

APPENDIX M

WAVE FREQUENCY AND MAXIMUM HEIGHT

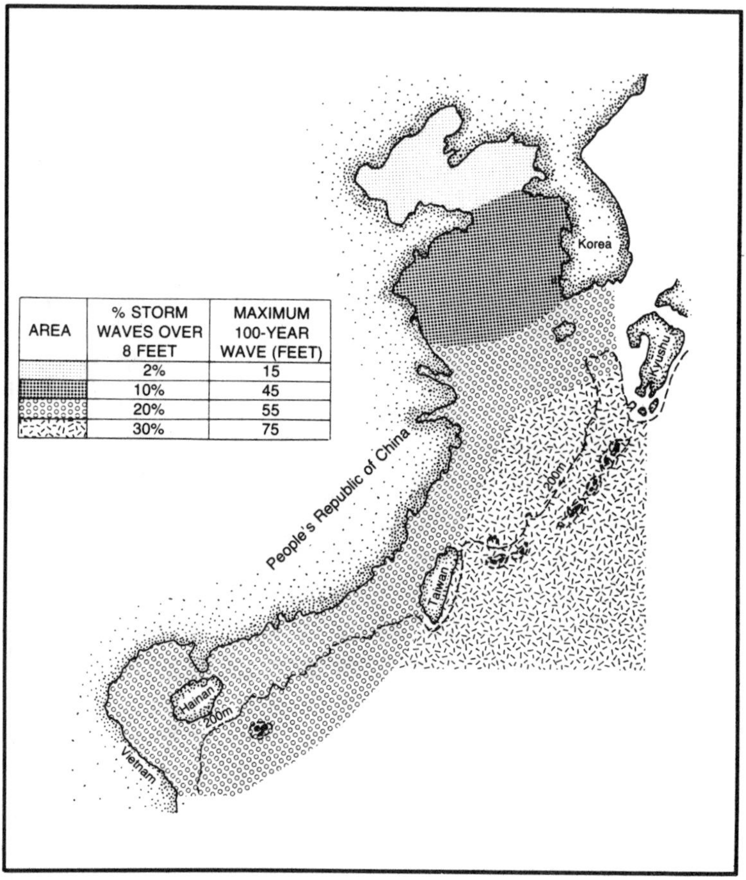

AREA	% STORM WAVES OVER 8 FEET	MAXIMUM 100-YEAR WAVE (FEET)
	2%	15
	10%	45
	20%	55
	30%	75

Source: Willums, *"China's Offshore Petroleum," The China Business Review* 4(14):13 (July-August 1977).

APPENDIX N

ENVIRONMENTAL/TECHNOLOGY CLASSIFICATIONS OF CHINA'S CONTINENTAL SHELF

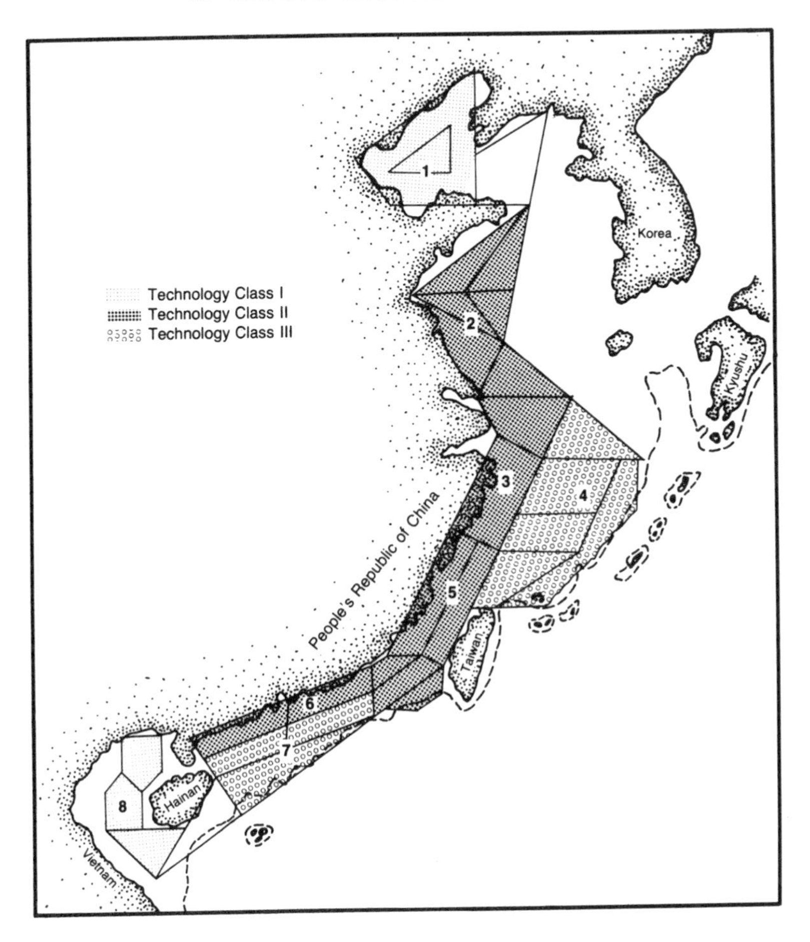

Source: Jan-Olaf Willums, *China's Offshore Oil* (unpublished Ph.D. diss.), 1975.

APPENDIX O

PRC ENERGY CONSUMPTION BY SECTOR
(Thousand bpd of Oil Equivalent)

Year	Total	Industry and Construction	Agri-culture	Transpor-tation	Residential & Commercial
1952	560	147	Negl.	66	347
1957	1,293	480	13	120	680
1965	2,467	1,187	80	187	1,013
1970	3,707	2,120	160	213	1,213
1974	5,067	3,133	320	253	1,360
1980:					
High	9,667	6,520	960	360	1,827
Medium	8,933	6,026	840	333	1,733
Low	8,213	5,547	733	320	1,613
1952	100%	26%	Negl.	12%	62%
1957	100	37	1%	9	53
1965	100	48	3	8	41
1970	100	57	4	6	33
1974	100	62	6	5	27
1980:					
High	100	67	10	4	19
Medium	100	67	9	4	20
Low	100%	68%	9%	4%	19%

Source: CIA, *Energy Balance Projections*, *op. cit.*, pp. 7, 33.

Totals may be off slightly due to rounding.

APPENDIX P

CALCULATED CAPITAL INVESTMENT INCREASES BETWEEN 1975 and 1980

As shown below, these percentages were calculated by projecting GNP in 1978 (the approximate midpoint of 1976-1980) to be 3.35 times that of 1955 (midpoint of 1953-1957), and assuming that the investment/GNP ratios for the two periods were approximately the same. Thus, 10.1 billion yuan would be about 5-6% of the total investment in 1978—the result of dividing 5.3 by 3.35 and multiplying that result by 3.5%, the share of total investment allocated to petroleum in 1953-1957. Performing a similar calculation for 1973 (as the midpoint for 1971-1975), the investment required to raise output from 36.7 MT (0.74 MMBD) in 1971 to 80 MT (1.6 MMBD) in 1975 (22% annual growth rate) was 4.3 billion yuan at 100 yuan per ton. This amounts to roughly 3% of national investment if we make the same assumptions as above and calculate GNP in 1973 at 2.64 times the 1955 figure.

CALCULATIONS:	1955	1973	1978
1. GNP index (1957 = 100):	87.15	230.8	291.4

2. 1978 GNP derived by projecting 1974 GNP at 5.2% annual growth rate (Ashbrook, p. 44).
3. 1978 Calculations:
   a) 291.4 (1978 projected GNP) ÷ 87 (1955 GNP) = 3.35.
   b) Projected investment from 1976-1980 = 200 MT (1980 crude output if growth at 20%/year) - 99 MT (1976 output with same growth rate) x 100 yuan/ton = 10.1 billion yuan.
   c) 10.1 ÷ 1.9 (investment from 1976-1980/that from 1953-1957) = 5.3.
   d) 5.3 ÷ 3.35 (ratio investment increase/ratio of GNP increase) = 1.6.
   e) 1.6 x 3.5 (relative increase x oil pctge. of total investment, 1953-57) = 5.6%.
4. 1973 Calculations:
   a) 230.8 (1973 GNP) ÷ 87 (1955 GNP) = 2.64.
   b) 80 MT (1975 output) - 36.7 MT (1971 output) x 100 yuan/ton = 4.3 billion yuan.
   c) 4.3 ÷ 1.9 (investment from 1971-1975/that from 1953-1957) = 2.26.
   d) 2.26 ÷ 2.64 (ratio of investment increase/ratio of GNP increase) = 0.86.
   e) 0.86 x 3.5 (relative decrease x oil percentage of total investment, 1953-1957) = 3.0%

   Conclusion: From 1971-75 oil commanded an average of 3% of total capital investment resources; from 1976-80 (to reach 4 MMBD in 1980), that share must average 5.6%, or nearly double.

Key Points in the Above Analysis:

a) The 100 yuan/ton figure is largely illustrative. Since it remains constant throughout the ratio series, its magnitude does not affect the final percentage increases.

b) Two assumptions are made:
   1) The capital/output ratio will remain roughly constant from 1971-1980.
   2) The investment/GNP ratios for 1953-1957, 1971-1975, and 1976-1980 are approximately equal.

c) Neither of the final oil investment %s contains any allowance for outlays to offset declines in production from older wells.

APPENDIX Q

CALCULATED CAPITAL INVESTMENT INCREASES BETWEEN 1975 and 1980 (Different Base Periods)

Series calculations for 1973-1977 (1975 midpoint) vs. 1978-1982 (1980 midpoint):

1. GNP index (1957 - 100):

1955	1975	1980
87.15	250.3	322.5

2. 1980 Calculations:
   a) 322.5 (1980 projected GNP) ÷ 87 (1955 GNP) = 3.7.
   b) Projected investment from 1978-1982 = 286 MT - 138 MT (1978 output with same growth rate) x 100 yuan/ton = 14.8 billion yuan.
   c) 14.8 ÷ 1.9 (investment from 1978-1982/investment from 1953-1957) = 7.8.
   d) 7.8 ÷ 3.7 (ratio of investment increase/ratio of GNP increase) = 2.11.
   e) 2.11 x 3.5 (relative increase x oil percentage of total investment, 1953-1957) = 7.4%.

3. 1975 Calculations:
   a) 250.3 (1975 GNP) ÷ 87 (1955 GNP) = 2.87.
   b) Projected investment from 1973-1977 = 115 MT (1977 projected crude output if growth at 20%/year) - 54.5 MT (1973 actual crude output) x 100 yuan/ton = 6.05 billion yuan.
   c) 6.05 ÷ 1.9 (investment from 1973-1977/investment from 1953-1957) = 3.18.
   d) 3.18 ÷ 2.87 (ratio of investment increase/ratio of GNP increase) = 1.11.
   e) 1.11 x 3.5 (relative increase x oil percentage of total investment, 1953-1957) = 3.9%.

APPENDIX R

PRC PETROLEUM INDUSTRY ORGANIZATION: STANDING COMMITTEE STATE COUNCIL

CHINESE ACADEMY OF SCIENCE

RESEARCH INSTITUTES (28)

MINISTRY OF PETROLEUM AND CHEMICAL INDUSTRIES

Responsible for discovery, production and petroleum refining, and downstream petrochemical production. Formerly Ministry of Fuel and Chemical Industries.

MINISTRY OF FOREIGN TRADE

FIRST MINISTRY OF MACHINE BUILDING

Responsible for the construction of the equipment for the petroleum industry.

CHINA NATIONAL CHEMICALS IMPORT AND IMPORT CORP.

SINOCHEM: Handles oil exports.

CHINA NATIONAL MACHINERY IMPORT AND EXPORT CORP.

MACHIMPEX: Responsible for petroleum machinery imports.

CHINA NATIONAL TECHNICAL IMPORT CORPORATION

TECHIMPORT: Department 2. In charge of importation of complete plants, new technology and technical patents.

SPECIAL AGENCIES OF THE STATE COUNCIL:

BUREAU OF SEISMOLOGY

Does oil exploration work.

BUREAU OF GEOLOGY

BUREAU OF OCEANOGRAPHY

Involved in offshore drilling.

Source: *The Chinese Petroleum Industry, op. cit.,* p. 22. © The National Council for U.S.- China Trade, 1976.

APPENDIX S

PETROLEUM-RELATED EQUIPMENT SALES TO CHINA

Item	Firm	Date or (Date Reported)	Value (US $ millions)
Seismic Surveying and Petroleum Prospecting Equipment			
Seismic Surveying Computer Equipment, including 3 seismic data processing sets each of which contained a model 704 Raytheon computer.	GeoSpace	10.73	5.5
Data Processing Center for Offshore Seismic Data Collecting and Prospecting, including 2 Control Data Cyber 172 Computer Systems (Medium Scale).	Compagnie Generale Geophysique (CGG)/Control Data Corp.	9.74	7.0
Comprehensive well-logging system with computer component and intermediate software supplied by Interdata of New Jersey.	US Firm	11.75	23.0
Seismic Exploration "Respond Systems" (2) with PDP 11/45 minicomputers and ancillary equipment.	Digital Resources Corporation	11.76	1.7
Seismic Monitoring Trucks (3) for interpreting data collected through mile-long cables with geological recording devices.	US Firm	12.76	1.5
		Subtotal	**38.7**
Down Hole Equipment			
Oil Drilling Equipment	France	1965	NVG
Steel Pipe	Mannesmann AG and Lowey Eng.	1967	NVG
Drilling Equipment and Spare Parts	US	12.73	0.5
Land Blowout Preventor Stacks (20)	Rucker	12.73	2.0
High-Pressure Hydraulic Fracturing Pumping Equipment	Dowell-Schlumberger	7.75	3.0
Oil Filters	San-netu	7.75	NVG
Well Heads and Gate Valves	Cameron Iron Works	11.75	0.75
Steel Casing for Drilling	Canada	11.75	0.7
Well Heads and Gate Valves	US Firm	11.75	1.0
Oil Well Servicing Equipment on Trucks (20)	Stewart and Stevenson Services, Inc.	2.76 5.76	5.6
		Subtotal	**13.55+**
Offshore Drilling Equipment			
Offshore Rigs and Spares	Romania	1966	NVG
Offshore Drilling Rig, *Fuji*, and 400-ton work-ship, *Kuroshio*	Offshore Drilling Co. of Japan	1972	8.4

Continued:

APPENDIX S (Continued)

Item	Firm	Date or (Date Reported)	Value (US $ millions)
Trailer Suction Hopper (4) Dredges	N.U. Industrieele Handels Combinate	1973	39.4
Self-propelling Bucket Dredgers (8)	Nippon/Kokan	6.74	53.0
Diesel Supply Boats, 660-Ton Capacity	Hitachi Shipbuilding Engineering	9.73	10.0
Oil Rig/Supply & Towing Vessels (8), 160 ft. long; DWT = 739.	Weco Shipping, Aarhus Flydedok A/S	late 73	20.0
"Jacket Type" Undersea Drilling Unit	Offshore Drilling (Teikoku Oil)	1973	NVG
Offshore Drilling Platforms (2)	JOD (Mitsubishi)	1973	41.0
Rig Oil Supply Ships (5)	Hitachi Zosen	12.74	NVG
Jack-up Oil Drilling Platforms (2)	Robin Loh Livingston Shipyard	5.75	60.0
Ocean Survey Ship	Japan Ocean Industry and Sumitomi Shoji	1975	NVG
Offshore Drilling Platform #2 Hakuryu (Heavy Duty)	Mitsubishi Heavy Industries	12.73	22.6
Tugboats, 9,000 HP, Pulling Capacity—82 Tons (2)	—	12.73	16.7
Jack-up rig	Hitachi-Loh	7.77	30.0
		Subtotal	**301.1+**
Other Petroleum Production Equipment			
Mechanical pumps for LPG (26)	Nishijima Manufacturing	12.64	NVG
Instruments	Stanhope-Seta Ltd.	5.66	NVG
Oil Lubricating Test Rigs (20)	Redman Heenam Froude	8.73	NVG
Onshore Oil Pipeline	Nippon/Kokan	1973	NVG
Pipelayers (12)	Komatsu	1974	1.0
Pipeline Tape, Coating and Wrapping Machines, etc.	Proline Pipe Equipment Ltd.	Spring 1975	NVG
38 Medium-sized Pipelayers & replacement parts.	Caterpillar Co.	10.75	3.8
Onshore Rigs (30?)	Romania	1975	NVG
Life Saving Systems Accessory to Oil Extraction Equipment	Nichimen	1975	NVG
Oil Field Transport Trucks for Moving Drilling Rigs	International Harvester	5.77	2.5
		Subtotal	**7.3+**
		TOTAL	**360.65+**

Source: *The China Business Review,* 4(4):10-11, July-August, 1977. © The National Council for U.S.-China Trade, 1977.

APPENDIX T

US TECHNICIANS IN CHINA

Company	Product	Number	Date	Location
Baker Trading Co.	Petroleum equipment	2	October 1975	Peking, Taching
		2	May 1976	Peking
		7	Feb.-March 1977	Peking, Szechuan
		6*	March-Apr. 1977	Peking & unknown oilfield
Boeing	707 aircraft, spare parts, ground equipment	6	Aug.-Sept. 1973	Shanghai
		4	Aug.-Oct. 1973 Oct. 1973-May 1974	Shanghai, then Peking
		2+	Aug. 1973-Dec. 1975	Shanghai, then Peking
Vendor: **Litton**		1	1974 (4 months)	Peking
Bucyrus-Erie	Blast hole drills, power shovels	3	1975 (3 months each)	Anshan Mine, Da Gou Mine, Dong Asham Mine, Chida Shan Mine, Nine Feng Mine, Wei Tou Shan, Pen-hsi Iron and Steel Works
		2	Feb.-Mar. 1977	Peking and unknown mines
Vendor: **General Electric**		1	Nov. 1975	Pen-hsi and others
Caterpillar	Pipelayers	5	Early 1976	Peking, Shengli
Control Data	Computer	7*	Aug. 1977 (4 wks.)	Peking
		2	Aug. 1977 (2 yr. shifts)	
Digital Resources	Seismic prospecting equipment	Not available	Late 1977	Unknown
Dowell Schlumberger	Oilfield equipment	1	Spring 1977	Peking
Engelhard (for Technospeichem)	Xylene process	1	1974	Taching
Geospace	Seismic surveying equipment	16*	May-June 1975	Peking
Lummus	Basic technology for Toyo ethylene plant	5	May 1976	Peking Petrochemical Works
Pullman Kellogg	Ammonia fertilizer plants	140 +	Nov. 1974-early 1978	Szechuan, Hunan, Heilungkiang, Liaoning, Hupei, Yunnan, Kweichow, Hopel.
Vendors:				
Babcock and Wilcox		6	1975-1977	
Benfield		3	1975-1976	
Bently Nevada	Vibration & monitoring instruments for turbines	2	July-Oct. 1976	
		1	Dec. 1976-Feb. 1977	
Betz		4	1975-1976	
C and CI		2	1975-1976	
CB & I (Chicago Bridge and Iron)		2	1975-1976	

APPENDIX T (Continued)

Company	Product	Number	Date	Location
Delaval		10	July, 1975-1977	
Dresser Clark		7	1975-1977	
Halliburton		4	1975-1977	
Honeywell		5	1975-1977	
IPM (UPS) System		1	1975-1976	
Los Angeles Water Treatment (a division of Chromalloy)	Water treatment equipment	5	Mid 1975-Early 1978	
Marley		1	1975	
OCI		2	1975-1976	
Westinghouse		1	1975-1976	
RCA Global Communications	Satellite Communications earth stations	23 25 + 35	Feb.-March. 1972 July-Sep. 1973 Oct. 1972-July 1973	Shanghai Shanghai Peking
Vendors:				Shanghai, Peking
Comtech	Electronic equipment			Shanghai, Peking
Cosmo	Consultant			Shanghai, Peking
E-Systems	Communications equipment			Shanghai, Peking
Rentronics	Tracking			Shanghai, Peking
Northern Electric	Multiplex equipment			Shanghai, Peking
United Technologies (Pratt & Whitney Aircraft Division)	Jet engines	2 +	Sept. 1973-Sept. 1974 (and other trips from Hong Kong)	Peking
WABCO	Mining trucks	4 +	March-Apr. 1976	Pen-hsi Iron and Steel Works
			March-July, Sept.-Dec. 1976, Feb. 1977	
Vendor:				
Cummins Engine		4 1	July 1975 Feb. 1977	Peking Pen-hsi
Western Union International	Satellite earth communications station	3 2 +	Mar. 1973 Aug. 1973-Apr. 1974	Peking Peking
Vendors:				
Comtech		5* 1	Oct. 1973-Jan. 1974 Aug. 1973-Oct. 1973	Peking Peking
E-Systems		5 1	Oct. 1973-Jan. 1974 Oct. 1973	Peking Peking
GTE		1	Mar. 1973	Peking

APPENDIX T (Continued)

Company	Product	Number	Date	Location
Caterpillar Far East Ltd.	Diesel and gas engines	2	August, 1977 (3 weeks)	Peking
Dowell	Petroleum equipment used in high-pressure hydraulic fracturing	1	April 26, 1977- May 20, 1977	Peking, Takang, Chungking
Dunegan-Endevco	Advanced acoustic emission instrumentation	1	Early October, 1977	Peking
		4	Late October, 1977 (1 week)	Peking
GE Westinghouse Caterpillar Ingersoll-Rand Clark Armco	Serviced Robin Loh Oil drilling rig for which all had supplied equipment	14	September, 1977	Nanning, Kwangsi Chuang
International Harvester	Oil field transport equipment	5-7?	January-March, 1978?	Unknown
Reed Tool	Drill bits	1	1975 (1 month)	Peking
Rolligon	Rough terrain vehicles for oil field use	1	April, 1978 (1 month)	Unknown
Smith International	Drilling bits	6	May, 1977 (1 month for part; 2 weeks for part)	Peking
Stewart & Stevenson	Oil well servicing equipment	4	March-April, 1977; June-September, 1977	Chungking, Peking
Texas Instruments	Seismic monitoring equipment	1	July, 1977 (3 weeks)	Peking, Nanking
		2	August, 1977 (4 weeks)	
		2	September-October, 1977	
Other Companies		146	1977-1979	
	Total: At least	577		

* Approximately

† Includes some wives and/or children

Note: list is not inclusive

Source: *The China Business Review,* 4(1):28-29, January-February, 1977, and 4(6):33, November-December, 1977. © The National Council for U.S.-China Trade, 1977.

APPENDIX U

SAMPLE SINO-JAPANESE CRUDE OIL CONTRACT, 1974

This Contract is made between China National Chemicals Import and Export Corporation General Headquarters (hereinafter called Seller) and (   )(hereinafter called Buyer), as a treaty of comity and friendship, upon the terms and subject to the conditions hereinafter set forth.

1. **Product Name**: Taching Crude Oil
2. **Quantity**:
   ***,000 L/T (1 L/T equals 1.016 Kg., and the crude oil weight conversion is handled according to SYB2206-60).
3. **Typical Inspection**:

	Factor	Test Method
1) Specific Gravity D 20/40°C	0.85-0.856	SYB2206-60
2) Sulfur Content wt. %	0.11 max.	GB387-64
3) Water Content wt. %	0.5 max.	GB260-64

4. **Price**:
   1) As to *00,000 L/T in the 1974 former half, Yuan *****/ bulk net weight L/T. (Yuan *****) FOB ex Dairen Port.
   2) As to *00,000 L/T in the 1974 latter half, July thru Dec., the negotiation is to be held separately in June, 1974, based on the prevailing price fluctuation that time in the international market.
   All deliveries shall be deemed to be complete when the crude oil has reached the flange connecting the delivery facilities ashore with the receiving facilities aboard the tanker. The crude oil shall be pumped at the risk and peril of Seller up to that flange only and thereafter title shall pass to Buyer with the resulting risk and peril.
5. **Deliveries**:
   *00,000 L/T in the 2nd quarter and *00,000 L/T in the 3rd and 4th quarters, the concrete deliveries setup shall be fixed separately.
   The delivery volume in each quarter shall be leveled evenly in the months.
   The request of an advance–or a deferred shipment of each quarter shall be advised for its consultation by the requesting party to the opposite party in advance of thirty days of the said quarter.

6. **Delivery Port:** Port Dairen, China
7. **Destination:** Ports in Japan
8. **Payment:**
   In accordance with Article 10, Item 2, of this Agreement, Buyer shall open the irrevocable, transferrable, divisible and payable in Yuan at sight letter of credit for The Chemicals Corporation, Dairen Division as the beneficiary, in advance of ten days before the lot loading as stipulated in the shipping volume schedule & due dates agreement, via a mutually concurred bank channel.
   This letter of credit shall be payable at the bank with the bill drawn by the beneficiary designating the opened bank as the payer, endorsed with the documents specified in Agreement, Article 9.
   The amount quoted in L/C shall be fixed respectively, allowable for ± 5% deviations, in accordance with the lots in the delivery schedule. The bank costs outside of China shall be borne by Buyer.
9. **Documents:**
   1) Seller shall submit the following documents to the payer bank in order to endorse the payment.
      a) Invoice–4 copies
      b) Clean Bill of Lading–1 original
      c) Inspector's Certificates of Product–1 copy each Quality, Weight & Place of Origin
   2) Seller shall forward two sets of the above-mentioned documents copy via air mail to Buyer within five days after tanker's departure.
   3) Seller shall submit two sets of the above-mentioned copies to the loading tanker in order to hand them over to consignee at delivery port.
10. **Loading Notice & Conditions:**
    1) The dirty ballast of Buyer's tanker shall be disposed by itself in advance of its arrival in the loading port. No waste water shall be emitted in the harbors.
    2) Buyer shall advise Seller of his offtake schedule by lots (specifying each tanker's name) in the following month, by cable in advance of twenty days before the said month. Seller shall answer to the Buyer's notice by cable within

five days of his acceptance.
Buyer shall make arrangements for the tanker schedule to the loading port in accordance with the mutually concurred offtake lots and tanker schedule.

3) Buyer shall give Seller (inclusive of its Port Dairen Regional Office) and to the Port Dairen export agent five days prior cable notice of deliveries required, specifying the name of vessel, its nationality, approximate date of delivery and quantity. Buyer shall give forty-eight hours advance notice before the port entry of his confirmation of what has been advised before.
4) At the Buyer's failure to get the assigned tanker to the loading port at the due date, the time lag in excess of one day shall be counted out in the running hours.
5) The draft of Buyer's tanker shall not come in excess of 31 feet at the peak load; the length shall be less than 600 feet; the loading capacity shall be less than 20,000 L/T. Any deviation from the above specifications shall be consulted fully with Buyer for his concurrence in advance, or the resulting dead space and the relevant loss is to be borne by Buyer.
6) After the loading is over, Seller shall notify Buyer by cable in twenty-four hours of the cargo's contract serial no., product name, specific gravity, vessel's name, consignee, loaded quantity, destination, invoice price and departure date.

11. Loading Regulations:

1) At the tanker's arrival in the loading port, the tanker captain take proceedings for the entry (inclusive of joint inspection, tank check, etc.) at the Port agency. If the proceedings are over and the loading conditions are all set, the captain hands over the loading conditions completion notice to the Port agent and, after signing his name for confirmation, sets about the counting-up in accordance with the stipulations as follows. The captain's notice shall be delivered in person within the office hours. If it is delivered by 12:00, the running hours count from 14:00; if delivered by 18:00, they count from 08:00 the following day. When the following day falls on a holiday, counting-

up is postponed till the first working day following.

2) The Seller's running hours shall be completed within the consecutive forty-eight hours, except for the case where a rough weather such as a gale and a thunderstorm stands in the way or where the Port Authorities stipulate otherwise. (The disconnection of its loading pipe shall be deemed as the loading completion time). The loading on a holiday is carried out at the Seller's option and is counted into the running hours. If loading is suspended on a holiday, the elapsed time is counted out of running hours.
3) After the loading is completed, Buyer shall allow for four hours in the Seller's preparation of loaded quantity conversion and the shipping documents. The grace hours are counted out of the running hours.
4) The elapsed lay time caused by the Port Authority stood in the way of moorage from its safety point of view shall be counted out of the running hours.
5) At the failure of the loading completion within the stipulated hours in this Article, Item 2, Seller shall pay the demurrage, to the Buyer in accordance with the charter party. The demurrage, however, shall not come in excess of the demurrage rate in the World Scale. The demurrage caused by the accident of machinery or power while in loading shall be reduced by half at settlement between Seller and Buyer.
6) The tanker's doings in the loading port shall be prepared by the Port agent in rigid conformity with the captain's running hours calculation endorsed by his signature.

12. Product Inspection:

1) Weight Measurement:

   The weight shall be quoted from the weight measurement certificate issued by the Product Inspection Bureau of China at the loading port; the weight measurement entered in the shipping documents shall be quoted from the Certificate. The quantity entered in the weight measurement as well as in the shipping documents shall be referred to by both Seller and Buyer as the delivery volume under this Agreement.
2) Quality Check:

   The product collected out of the bulk by the Product Inspection Bureau in accordance with the sampling procedure if SYB2001-59 is, after blending, divided into four samples as the standard reference of the Seller's product. The product quality certificate issued by the Product Inspection Bureau after the analysis of one of the samples is the basic reference of the Seller's product. Out of the rest, one sample is entrusted with the tanker for the Buyer; one is retained by the Seller; one is retained by the Product Inspection Bureau for retest and arbitration.

3) When the product arrives at the destination, Buyer shall have the right to re-check the product's quality and quantity on the Buyer's account. At any discrepancy in quality and quantity with the contract stipulations, Buyer shall have the right to make a claim for the damages. The natural loss during the course of transportation and the quality/quantity change during the same while, however, shall not be covered under the damages. The claim shall take effect if it is raised within thirty days after the product has arrived at the destination.

13. **Force Majeure**:

At the Seller's failure by dint of Force Majeure to make the delivery by the due date, Seller shall have the option either to defer the delivery, partly or wholly, or to withdraw the delivery under contract, partly or wholly. Seller shall be obliged to submit to Buyer the exhibit prepared by the China Council for the Promotion of International Trade in order to attest the occurrence of such an accident.

14. **Penalty**:

At the Buyer's failure to open a letter of credit in spite of the stipulations in the Agreement, Buyer shall be imposed of a penalty for the Seller's damages caused by the resulting hindrance to fulfill the due delivery, partly or wholly. The penalty amount shall be 1% of the total amount under contract or the amount equivalent to the hampered delivery, case by case.

15. **Arbitration**:

At any and all the disputes caused relevantly or resultingly with this Agreement, both Buyer and the Seller shall be obliged to consult with each other for the amicable settlement. If Buyer and Seller can not reach a solution of the issue through the consultation, either of the parties can submit the issue to arbitration, which shall be lodged at the competent person in the respondent's country. If the arbitration is lodged in China, the arbitrator shall be the International Trade Promotion Committee, Foreign Trade Arbitration Committee who shall handle the business in accordance with its Regulations & Arbitration Procedure. If it is lodged in Japan, the International Commerce Arbitration Association shall handle it in accordance with its Arbitration Procedure and Regulations; the arbitrator shall not necessarily be selected out of the Arbitrator Register but be limited to those of the Chinese or the Japanese nationality or of other nationalities concurred by both of Seller and Buyer. The judgment of the arbitration shall be deemed final, forcing both of Seller and Buyer under the Agreement to act on it. Both Seller and Buyer shall give every facility as well as the guaranteed security, upon the endorsement of each of the respective Government, to the arbitrator who will frequent the countries of Seller and Buyer. The arbitration cost shall be borne by the case loser, except for the case where the arbitrator sets forth otherwise.

16. This Agreement is made both in Chinese and Japanese and both of them shall take effect equally.

Source: China's Petroleum Industry, Spec. Rep. No. 16, *op. cit.*, pp. 116-17.

APPENDIX V

JAPAN'S FIVE-YEAR SUPPLY AND DEMAND PLAN
(Thousand bpd)

	1976	1977	% Chg. '77/'76	1978	% Chg. '78/'77	1979	% Chg. '79/'78	1980	% Chg. '80/'79	1981	% Chg. '81/'80
Total Demand	4,248	4,515	+ 6.3	4,690	+ 3.9	4,901	+ 4.5	5,155	+ 5.2	5,435	+ 5.4
Crude Imports	4,762	4,993	+ 4.9	5,239	+ 4.9	5,473	+ 4.5	5,678	+ 3.7	5,996	+ 5.6
Product Imports	525	530	+ 1.0	529	- 0.2	534	+ 0.9	555	+ 3.9	561	+ 1.1
Product Exports	343	383	+11.7	407	+ 6.3	429	+ 5.4	452	+ 5.4	476	+ 5.3
Inland Demand	3,906	4,134	+ 5.8	4,282	+ 3.6	4,470	+ 4.4	4,702	+ 5.2	4,958	+ 5.4
Gasoline	524	548	+ 4.6	571	+ 4.2	595	+ 4.2	617	+ 3.7	638	+ 3.4
Naptha	607	682	+12.4	718	+ 5.2	762	+ 6.1	805	+ 5.6	848	+ 5.3
Kerosene	426	433	+ 1.6	458	+ 5.8	485	+ 5.9	513	+ 5.8	541	+ 5.5
Gas Oil	291	305	+ 4.8	320	+ 4.9	334	+ 4.4	348	+ 4.2	361	+ 3.7
Fuel A	349	368	+ 5.4	391	+ 6.3	414	+ 5.9	438	+ 5.8	461	+ 5.3
Fuel B	162	153	- 5.6	153	0.0	154	+ 0.7	154	0.0	154	0.0
Fuel C	1,509	1,606	+ 6.4	1,630	+ 1.5	1,683	+ 3.3	1,782	+ 5.9	1,905	+ 6.9

Sources: *Petroleum Intelligence Weekly*, April 18, 1977, pp. 5-6; *Platt's Oilgram Service*, March 29, 1977, p. 4, and April 1, 1977, p. 3.

APPENDIX W

VARIOUS PROJECTIONS OF CHINESE AND SOVIET OIL OUTPUT

	CHINA Cheng[1]		CHINA Lewis[2]		CHINA Kambara[3]		CHINA Park/Cohen[4]		CHINA Harrison[5]	
Year	Output (MMBD)	Growth Rate (%)	Output (MMBD)	Growth Rate (%)	Output (MMBD)	Growth Rate (%)	Output (MMBD)	Growth Rate (%)	Output (MMBD)	Growth Rate (%)
1974	1.26	--	1.30	--	1.04	--	1.30	21	1.30	--
1975	1.52	20	1.57	21	1.36	30	1.57	21	1.56	20
1976	1.82	20	1.90	21	1.62	20	1.90	21	1.79	15
1977	2.18	20	2.30	21	1.94	20	2.30	21	2.06	15
1978	2.56	17	2.79	21	2.32	20	2.78	21	2.37	15
1979	3.00	17	3.37	21	2.66	15	3.36	21	2.73	15
1980	3.52	17	4.08	21	3.06	15	4.06	21	3.14	15
1981	4.04	15	4.94	21	3.36	10			3.45	10
1982	4.64	15	5.98	21	3.70	10			3.80	10
1983	5.34	15	7.23	21					4.18	10
1984	5.98	12	8.75	21					4.60	10
1985	6.70	12	10.59	21					5.06	10
1986									5.57	10
1987									6.13	10
1988									6.74	10
1989									7.41	10
1990									8.15	10

1. Cheng, *China's Petroleum Industry*, *op. cit.*, p. 39.
2. Christopher Lewis, "Outlook Bright for Oil, Coal," *FEER*, October 4, 1974, p. 27.
3. Kambara, "The Petroleum Industry in China," *op. cit.*, p. 717.
4. Park and Cohen, "China's Oil Weapon," *op. cit.*, p. 40. Only 1980 production figure cited. Interim years filled in based on required annual increases to reach 4 MMBD by 1980.
5. Harrison, "Time Bomb in East Asia," *op. cit.*, p. 4. Only production figure of 8 MMBD by 1989-1990 cited. Interim years projected based on conversation with the author and yearly growth necessary to achieve the 1989-1990 target. The author spells out the assumptions underlying the 1990 projection in his recent book, *China, Oil and Asia*, *op. cit.* The book does not seek to project growth rates on a year-by-year basis, but treats average annual growth of 12 - 13% over the 1975-1990 period as likely in the context of the assumptions set forth.

APPENDIX W (Continued)

VARIOUS PROJECTIONS OF CHINESE AND SOVIET OIL OUTPUT

	CHINA		CHINA		CHINA		CHINA		USSR		
	IDA[6]		Takahashi[7]		Riggin[8]		China Projections Based on USSR Analog[9]		Actual Output of USSR[10]		
Year	Output (MMBD)	Growth Rate (%)	Output (MMBD)	Growth Rate (%)	Output (MMBD)	Growth Rate (%)	Output (MMBD)	Growth Rate (%)	Year	Output (MMBD)	Growth Rate (%)
1974	1.30	23	1.22	22	1.30	--	1.30	20 (Actual)	1955	1.42	20
1975	1.60	20	1.6.	32	1.63	25	1.60	20 (Actual)	1956	1.68	18
1976	1.92	20	2.06	28	2.04	25	1.82	13.7	1957	1.97	17
1977	2.30	20	2.68	30	2.55	25	2.08	13.7	1958	2.26	15
1978	2.76	20	3.16	18	3.19	25	2.36	13.7	1959	2.59	14
1979	3.31	20	3.36	6	3.98	25	2.68	13.7	1960	2.96	14
1980	3.97	20			4.98	25	3.04	13.7	1961	3.32	12
1981							3.46	13.7	1962	3.72	12
1982							3.94	13.7	1963	4.12	11
1983							4.24	7.5	1964	4.47	9
1984							4.56	7.5	1965	4.86	9
1985							4.90	7.5	1966	5.30	9
1986							5.26	7.5	1967	5.76	9
1987							5.66	7.5	1968	6.18	7
1988							6.08	7.5	1969	6.57	6
1989							6.54	7.5	1970	7.05	7
1990							7.04	7.5	1971	7.44	5
									1972	7.88	6
									1973	8.58	9

6. Colm, *et al.*, *Chinese Petroleum Developments to 1980, op. cit.*, pp. 11, 48.
7. Shogoro Takahashi, *Nitchu Keizai Kaiho* (China-Japan Economic Report), December 1973, cited in Choon-ho Park, *Energy Policies of the World: China* (Newark, Del.: Center for the Study of Marine Policy, 1975), p. 27.
8. Charles H. Riggin, *Oil and the People's Republic of China: Status and Future Exploration of China's Oil Resources* (Maxwell AFB, Alabama: Air War College, April 1976), p. 25.
9. Analog constructed by projecting PRC growth from 1976 (1.6 MMBD) to 1982 (4 MMBD) at average 13.7% rate which Soviets expanded from 1956 (1.6 MMBD) to 1963 (4 MMBD). Process was repeated by projecting PRC growth from 1982 to 1990 at 7.5%, the average annual Soviet increase from 1963 (4 MMBD) to 1973 (8 MMBD).
10. Iain F. Elliot, *The Soviet Energy Balance* (N.Y.: Praeger Publishers, Inc., 1974), pp. 76-77; and Robert E. Ebel, *Communist Trade in Oil and Gas* (N.Y.: Praeger Publishers, Inc., 1970), p. 40.

Notes

Notes

1. See, for example, Choon-ho Park and Jerome A. Cohen, "The Politics of China's Oil Weapon," *Foreign Policy* 20:40 (Fall 1975).

2. Selig S. Harrison, "Time Bomb in East Asia," *Foreign Policy* 20:4 (Fall 1975). More recently, Harrison has qualified this estimate to indicate there is a "better than fifty-fifty chance" of reaching an 8 million bpd production level in 1990. See Selig S. Harrison, *China, Oil and Asia: Conflict Ahead?* (New York: Columbia University Press, 1977), pp. 19-20.

3. Chart is a composite from the following sources: *Platt's Oilgram Service,* 30 April 1975, p. 1, 2 December 1976, p. 1, and 9 November 1977, p. 3; U.S., Congress, Joint Economic Committee (JEC), a compendium of papers on *China: A Reassessment of the Economy,* article by Bobby A. Williams, "The Chinese Petroleum Industry: Growth and Prospects," 94th Cong., 1st sess., 10 July 1975, p. 228; *Far Eastern Economic Review* (hereafter *FEER*), 23 January 1976; *The China Business Review* 4(4):29, 32 (July-August 1977); *Petroleum Intelligence Weekly,* 21 January 1976, p. 6; *China's Petroleum Industry,* Special Report no. 16, The National Council for U.S.-China Trade (June 1976), p. 50;

Charles H. Riggin, *Oil and the People's Republic of China: Status and Future Exploration of China's Oil Resources* (Maxwell AFB, Ala.: Air War College, April 1976), pp. 70-73, 139; *New China News Agency,* 26 December 1977.

4. U.S., Central Intelligence Agency, *China: Energy Balance Projections,* A (ER), 75-75, November 1975, p. 1. A more recent CIA study, devoted exclusively to China's oil prospects, has revised the 1980 production and export figures for PRC oil to 2.4 to 2.8 million bpd and 200,000 to 600,000 bpd, respectively. U.S., CIA, *China: Oil Production Prospects,* ER 77-100304, June 1977, p. 1. Similar estimates have also been made by the Sino-British Trade Council. See John Cranfield, "Mainland China Gearing Up to Boost Oil Exports," *Oil and Gas Journal,* 11 August 1975, p. 24.

5. CIA, *Energy Balance Projections,* p. 17. Using its medium energy supply case, CIA estimates that annual GNP growth would be reduced from 7.3 to 5.8%.

6. Some of the information required before attempting a realistic reserve estimate: an idea of the "cutoff value" (i.e., whether it takes organic matter concentrations of less than 1, 1, 2, 3, or more percent to produce oil from a particular formation) and source bed location and thickness. Since wells are usually drilled on anomalies (i.e., individual traps), they seldom go deep enough to obtain reliable data on the net acre-feet of mature source beds necessary for total resource calculations.

7. The main marine or noncontinental basins: Taiwan/Kwangsi/Kweichow (Paleozoic platform) and Luichow-Taiwan/Kiangsu/northeast portion of North China basin (coastal basins with Tertiary marine sediments). A. A. Meyerhoff, "Developments in Mainland China, 1949-1968," *American Association of Petroleum Geologists Bulletin* 54 (8): 1574 (August 1970). The U.S. Geological Survey, in addition, believes that the thickest areas of sediments in Kiangsu (Yellow Sea and East China Sea) and in the

northeast portion of North China basin (Po Hai) have continental graben facies of earlier Tertiary age which are only overlaid by sediments, possibly marine, of the later Tertiary and Quaternary ages.

8. Interview with Maurice Terman et al., U.S. Geological Survey (USGS), Reston, Va., 3 February 1976. For a description of drilling difficulties in the Uinta basin, see L. D. Findley, "Why Uinta Basin Drilling Is Costly and Difficult," *World Oil* 174(5): 77-80, 92 (April 1972); Peter T. Lucas and James M. Drexler, "Altamont-Bluebell: A Major Fractured and Overpressured Stratigraphic Trap, Uinta Basin, Utah" (paper for 1975 symposium of Rocky Mountain Association of Geologists, Denver, Colo.). Both articles describe unusual reservoir conditions (vertically separated, highly fractured, low porosity-producing zones of variable lithology) and fluid properties (undersaturated, high pour-point oil) which do not lend themselves to normal exploration and production techniques. The result is a series of wells that are difficult to drill and expensive to operate, and whose ultimate recoveries and product quality are predictable only after a long production history. Findley calculates well-drilling and completion costs range from $400,000 to $2 million; capital outlays for lease production facilities vary from $75,000 to $150,000; monthly per-well operating expenses are $1,500 to $3,000. In addition, USGS discussions with visiting PRC geologists have generally confirmed validity of the Uinta basin analogy and exploration difficulties associated with the unique geology of China's three main fields.

9. A. A. Meyerhoff and Jan-Olaf Willums, "Petroleum Geology and Industry of the People's Republic of China," United Nations, Economic and Social Commission for Asia and the Pacific, Committee for Coordination of Joint Prospecting for Mineral Resources in Asian Offshore Areas *(CCOP) Technical Bulletin* 10: 176-181 (December 1976). Meyerhoff and Willums see Ta-ch'ing being comprised of

up to ten major fields, the largest three of which are located in the Ta-t'ung-chen structure near the center of the Sung-liao basin. Due to its enormous size and shallow depth, Ta-t'ung-chen appears to be something of an anomaly among Sun-liao basin fields. Not a basement horst or arch, it is instead located in the midst of a geologic environment where lacustrine (and therefore winnowing) conditions were persistent during deposition. Meyerhoff and Willums thus place the average per-well production of Ta-t'ung-chen fields at approximately 200 bpd, with that from other Sung-liao basin (Ta-ch'ing) fields being much less.

10. See Vaclav Smil, "Communist China's Oil Exports: A Critical Evaluation," *Issues and Studies,* March 1975, p. 3. Smil cites estimates ranging from 75 to 750 bpd. It seems likely, however, that the higher figure may include both oil and water produced from each well. In a December 1976 interview with Melvin Searles, then vice-president of the National Council for U.S.-China Trade, the author was informed that Chinese officials quoted an output figure of 520 to 740 bpd to a visiting delegation of the National Council. But the Chinese also indicated that this figure was roughly 50% water, putting actual crude oil production in the range of 260 to 370 bpd. Furthermore, none of the observer data distinguishes between production and injection wells and how those ratios might affect per-well production rates. Because of this uncertainty, some U.S. government sources tend toward 300 bpd as a more likely per-well figure.

11. For a description of exploration and production problems at Ta-Kang, see *Peking Review,* 24 May 1974.

12. For a more detailed discussion, see CIA, *China: Oil Production Prospects,* pp. 12-18.

13. As of early 1976, one estimate gave the following well statistics for some of China's fields. See Meyerhoff and Willums, *(CCOP) Technical Bulletin* 10: 121, 176, 180, 189, 191.

Oil Field	No. of Drilling Rigs	No. of Wells	No. of Well Locations
Ta-ch'ing	7	4,069	176
Sheng-li	81	1,406	348
Ta-kang	52	700	76
P'an-shan	46	229	104
K'o-la-ma-i (West China)	14	1,540	427

14. Most geologists regard much of the Po Hai as a lateral extension of onshore fields at Ta-kang and Sheng-li. If this is true, the above-mentioned structural difficulties, plus normal additional expenses associated with offshore drilling, would make full-scale production from that source extremely expensive.

15. For example, both Prudhoe Bay and the North Sea are characterized by predominantly stratigraphic trapping mechanisms.

16. Terman, USGS, 1976 interview.

17. *Baltimore Sun,* 7 January 1977; *New York Times,* 7 January 1977, p. 32.

18. Ross H. Munro, "Chinese Oil Boom in Ta-ching Slows," *Washington Post,* 19 December 1976. In the same vein, see Selig S. Harrison, "Petroleum Resources in the Western Pacific and Indian Oceans: The Potential for Conflict or Cooperation" (paper prepared for the Conference on the Environments for U.S. Naval Strategy in the Pacific–Indian Ocean Area, 1985-1995, at U.S. Naval War College, Newport, R.I., 24-26 March 1977), p. 8. Harrison states that his extensive interviews with geologists and petroleum engineers show the combined output of Ta-ch'ing, Sheng-li, and Ta-kang is not likely to reach 2 million bpd before 1986-1988 and is never going to exceed that amount greatly.

19. Meyerhoff and Willums, *(CCOP) Technical Bulletin* 10:181. The strong influence of Soviet technicians apparently dictated this practice to obtain immediate gains in production and to delay the installation of scarce pumping units used in normal extraction operations.

20. See comments of Necmettin Mungan in *China's Energy Policies and Resource Development,* compiled by Thomas Fingar (report of a seminar, 2-3 June 1976, Stanford University), p. 40. The line drive approach involves pumping water into the ground through a series of injection wells, which then forces oil to the surface through an adjacent series of extraction/production wells. After a short time, however, cusps begin to form in the unswept areas between extraction wells. Where the line drive technique uses two injection wells for every three extraction wells, five- and nine-spot water-flooding gives ratios of 1:3 and 1:5, respectively, with associated increases in overall sweep efficiency. The three processes can be diagramed as follows:

Line Drive	Five Spot Water Flood	Nine Spot Water Flood
X X X X X	0 0 0 0	0 0 0 0 0 0
0 0 0 0 0	X X X	0 X 0 X 0 X
0 0 0 0 0	0 0 0 0	0 0 0 0 0 0
0 0 0 0 0		
X X X X X		

X = Water injection wells.
0 = Production/extraction wells.

Mungan's observations were made during a visit to Ta-ch'ing in early 1976 as a petroleum consultant. His conversations with production personnel indicated that they were well aware of the potential engineering problems created by overdriving Ta-ch'ing but that they were often overruled by "political considerations."

21. For instance, refer to:

a. 49 BB (19.7 proved + 29.2 undiscovered), China's Production May Jump Sixfold by 1985," *World Oil,* October 1975, p. 154. This estimate was given in a paper by W. E. Humphrey (vice-president for Oil Exploration, Amoco International Oil Corp.) presented to Ninth World Petroleum Congress, Tokyo, July 1975.

b. 45 to 57 BB (onshore only), JEC, Williams, "The Chinese Petroleum Industry," p. 225.
c. 52 BB (onshore only), 1966 official PRC estimate quoted in *Petroleum Times,* 11 July 1975.
d. 70 to 75 BB, represented as the consensus of most "U.S. experts" in *New York Times,* January 1975, p. F-3.
e. 40 BB onshore and 40 BB offshore, cited by the CIA as the consensus of academics, oil companies, and USGS personnel in its study, *China: Oil Production Prospects,* p. 1.
f. 30 to 35 BB onshore and 10 to 15 BB offshore, M. Terman, USGS, cited in Fingar, *China's Energy Policies,* p. 32. Terman's estimate is based on his own tectonic mapping and use of the Uinta basin analogy. He acknowledges that offshore estimates could substantially increase if further exploration demonstrates East China Sea basins contain marine sediments instead of continental deposits.

22. Chinese oil production from 1965 to 1975 totaled about 4.8 BB. Assuming a 15 percent production growth rate from 1976 to 1980, additional usage would equal:

1976	-	710 MMB	1979	-	1.1 BB
1977	-	820 MMB	1980	-	1.3 BB
1978	-	940 MMB	Total 1976-1980	-	4.9 BB

Of this nearly 10 BB total from 1965 to 1980, 2 to 3 BB already represented production from interior fields or had occurred before most reserve estimates were made. The result is usage of 7 BB from time of reserve estimation until 1980 and usage of 12 to 13 BB (3.5 million bpd x 365 days/year x 10 years, just to maintain a baseline production level of 3.5 million bpd) from 1980 to 1990. This total usage of 20 BB equals the maximum likely reserve for eastern China, derived from optimistic assumptions about total PRC reserves (0.20 x 100 BB) portrayed in Appendix C.

23. The CIA, using slightly different production and reserve figures, observes the same phenomenon. CIA, *China: Oil Production Prospects,* p. 22. If the North/Northeast continues to produce 75 to 80% of PRC output, even a slowdown in petroleum growth to 10% per year will exhaust China's 50% reserve level (50% probability of the estimated quantity of petroleum referenced in the private study in Appendix C, 18.8 BB, will ultimately be recovered) for eastern fields by 1992 and its 0% reserve value for those fields by 1996. With higher annual oil growth, the time frame accelerates:

Growth Rate	Year of Exhaustion at 50% Reserve Level	Year of Exhaustion at 0% Reserve Level
15%	1989	1991
20%	1987	1989

In sum, even under favorable conditions (10% annual growth), eastern China's reserves will run out in 20 years. At higher growth rates, there is a strong possibility they may be exhausted in the next decade.

24. The classic example of this policy was Ta-ch'ing. Although discovered in 1959, it was not mentioned in the Chinese press until 1964 (at which time production totaled well over 100,000 bpd), and its exact location remained a secret until the early 1970s. It seems quite possible, however, that the secrecy surrounding such earlier oil discoveries was due in large part to their proximity to the USSR and to China's general ill feelings toward the Soviet Union after 1960.

25. Meyerhoff, "Developments in Mainland China," p. 1577; also, *U.S. News and World Report,* 9 December 1975.

26. China's exploratory drilling activity has reportedly increased some 400% over the 1950s, with Chinese crews now drilling 26,000 to 30,000 feet of hole each year. *World Oil,* October 1975, p. 153.

27. Interview with William Clarke, Bureau of East-West Trade, Domestic and International Business Administration, U.S. Department of Commerce, 21 July 1977.

28. The corollary, of course, is that deeper drilling might conceivably strike a large Prudhoe Bay–like structure.

29. See *Report of the Canadian Electrical Power Mission to the People's Republic of China, August 29–September 18, 1973,* Ottawa [n.d.].

30. It has been estimated that from 1965 to 1973 the Chinese achieved over 50% of their refinery expansion through such "technical transformation" at about 40% of the cost to build new plants. See JEC, Williams, "The Chinese Petroleum Industry," p. 245. Comparable figures for 1970 and 1971 are quoted in *Chung-kuo Hsin-win,* 3 September 1973, pp. 4-5.

31. Meyerhoff and Willums, *(CCOP) Technical Bulletin* 10:207. This goal is consistent with the average 15% annual growth in PRC refinery capacity over the past four years.

32. *China's Petroleum Industry,* pp. 37-39. China's synthetic fiber production in 1980 is expected to reach 500,000 metric tons, or roughly nine times 1975 output, as synthetics replace existing cotton production (which currently diverts needed agricultural resources away from grain production).

33. Chart is a composite estimate from following sources: *FEER,* 13 February 1976, pp. 74-75; *Journal of Commerce,* 12 January 1976, p. 28; Cranfield, "China Gearing Up," p. 24. Chu-yuan Cheng, *China's Petroleum Industry: Out-Put Growth and Export Potential* (New York: Praeger Publishers, Inc., 1976), pp. 102, 103; CIA, *China: Oil Production*

Prospects, pp. 12, 26; *Petroleum Intelligence Weekly,* 25 October 1976, p. 9; *Platt's Oilgram Service,* 1 December 1976, p. 4; *New China News Agency,* 22 July 1976; *U.S.-China Business Review* 2:54 (March-April 1976); V. Smil, "Communist China's Oil Exports Revisited," *Issues and Studies* 12 (9):71 (September 1976); *China's Petroleum Industry,* pp. 47-48. Exact port location shown on map in Appendix B.

34. *Hsin-hua,* 27 December 1975; *FEER,* 13 February 1976, p. 74. The exact mixture of oil, general cargo, and multipurpose berths among the 40 is not specified, but *Hsin-hua* implies that at least some oil wharves are in the 25,000 dwt range.

35. *Petroleum Intelligence Weekly,* 24 November 1975, p. 9; *FEER,* 13 February 1976, p. 75. The PRC had previously purchased a 75,000 dwt Norwegian tanker in late 1974. JEC, Williams, "The Chinese Petroleum Industry," p. 247.

36. Port limitations, however, do put the PRC at a competitive disadvantage with major exporters to Japan. See A. Wrightman, "Japan and China's Oil—Proceeding With Caution," *U.S.-China Business Review* 3(2):34 (March-April 1976). Most of Japan's crude imports from the Middle East come in 300,000 dwt tankers and those from Indonesia usually arrive in 150,000 dwt vessels. Imports from China are typically carried in 20,000 to 30,000 dwt ships. The result is roughly similar transportation costs ($0.70 to $0.80 per barrel) from the Persian Gulf to Japan (a 15,000-mile distance) as from Ta-lien to Tokyo (1,250 nautical miles).

37. CIA, *China: Oil Production Prospects,* p. 25. Chinese manufactured pipe over 12 inches is spiral formed and apparently unable to withstand high pressures. Although there are several recent reports of a 39-inch diameter pipeline from the south coast port of Chan-chiang to the refinery at Mao-ming, this pipe was also presumably imported. See *The China Business Review* 4(3):45, 54 (May-June 1977).

38. Listing represents a composite estimate from the following sources: *World Oil,* October 1975, p. 145;

Meyerhoff, "Developments in Mainland China," p. 1580; JEC, Williams, "The Chinese Petroleum Industry," p. 246; *Journal of Commerce,* 12 January 1976, p. 28; CIA, *China: Oil Production Prospects,* p. 25; *China's Petroleum Industry,* p. 43; *The China Business Review* 4(3):45, 54 (May-June 1977). For approximate routes of major pipelines, see Appendix B.

39. Some approximate distances from major interior basins to probable ports:

Szechwan/Chan-chiang	800 miles
Ordos/Tientsin	500 miles
Tsaidam/Tsingtao	1,400 miles
Dzungarian/Ch'in-huang-tao	1,600 miles
Tarim/Ch'in-huang-tao	1,800 miles

Besides generally rugged terrain, lack of many hard-surface roads, and few interior railroads (one single-gauge line stretching from Lan-chou to Urumchi represents the extent of present railroad construction west of Szechwan) would make the logistics of labor and materials transport to and from construction sites a tremendous task.

40. In addition to the detailed analysis in Appendix G, general reports from Japanese importers indicate that Ta-ch'ing crude has a 70.5% yield of residual oil, and Sheng-li has an 84% residual yield. *Platt's Oilgram Service,* 16 October 1975, p. 3. Similar sources also verify the high pour-point/high wax problems for both crudes and the water/sediment contamination for Sheng-li. See *FEER,* 28 November 1975, p. 47.

41. R. M. Smith, "China: The Next Oil Giant?" *Newsweek,* 27 October 1975, p. 39.

42. Quoted in Harrison, "Time Bomb in East Asia," pp. 7-8. This prediction has been criticized, however, since it simply calculated volumes of sediments without considering the tectonic framework of China's offshore regions.

43. U.S. oil company efforts have remained somewhat restricted by conflicting territorial claims, severe weather conditions, and a persistent suspicion that oil production in China's offshore regions will be gas limited.

44. See Terman, *China's Energy Policies,* p. 32; Meyerhoff and Willums, *(CCOP) Technical Bulletin* 10:194-200; and A. A. Meyerhoff, "Eastern Asian Coasts and Offshore Are Promising Petroleum Frontiers," *Oil and Gas Journal,* 27 December 1976, pp. 218-222. Although Terman estimates potential offshore reserves at 10 to 15 BB, he emphasizes that the complex, predominantly stratigraphic, traps of eastern China probably extend seaward, leading to such geologic difficulties that extensive exploitation of these structures is not likely to be economic. While Meyerhoff sees potentially recoverable resources of 30 BB, he also believes pre-Miocene objectives (those whose age make them the most promising oil targets) are limited, and that all large accumulations will be in Miocene-Pleistocene sediments, which constitute a much less attractive geologic environment. As evidence of this condition, Meyerhoff states that over 80 wells have been drilled in the Po Hai alone between 1969 and 1976, but that the erratic distribution of reservoirs has led to a high dry hole rate, even within productive structures.

45. Appendixes J and K are reproduced from exhibits by Jan-Olaf Willums in *China's Energy Policies,* pp. 51-52. They are based on Willums' interviews with geologists (Meyerhoff, Douglas Klemme of Weeks Associates, Dr. Papp of Chevron, and others) familiar with the East Asian offshore area and on the output of a mathematical model which calculates average field size and discovery decline curves, employing data from the interviews, through use of a Monte Carlo simulation technique.

46. For an excellent description of the issues and possible courses of action involved, see Harrison, *China, Oil and Asia.* In Harrison's words, greater PRC offshore activity will enable Peking to demonstrate the legitimacy of its territorial

claims with "survey ships and rigs instead of gunboats."

47. Columnist Joseph Kraft has reported Ta-kang wells within a few yards of the Po Hai. See *Washington Post,* 12 December 1975.

48. Appendix N comes from Jan-Olaf Willums, "China's Offshore Oil: Application of a Framework for Evaluating Oil and Gas Potentials Under Uncertainty" (Ph.D. diss., Massachusetts Institute of Technology, 1975), pp. 64, 286-303. Willums divides China's offshore areas into three environmental classes (I, mild; II, more difficult; III, frequently severe conditions) which correspond to the levels of offshore technology required to exploit them. Thus, Class I is a shallow, calm area, where it is relatively easy to drill; Class III is on a deep section of the continental shelf, subject to frequent storms and weather extremes, where drilling is extremely difficult and requires the best in offshore technology.

49. *Wall Street Journal,* 7 September 1977. The PRC is also reportedly interested in purchasing a second semisubmersible rig from Norway.

50. Chart is a composite from the following sources: *China's Petroleum Industry,* pp. 36, 45-46; *U.S.-China Business Review* 3(4):46 (July-August 1976); *Platt's Oilgram Service,* 5 January 1976, p. 4; 7 January 1977, p. 5; and 7 June 1977, p. 3; Harrison, "Resources in the Western Pacific," p. 22; Willums, "The Development of China's Petroleum Industry" (paper presented to the National Council on U.S.-China Trade's Conference on China's Oil Industry, Houston, Texas, 23 June 1976), p. 9; Meyerhoff and Willums, *(CCOP) Technical Bulletin* 10:206; *Wall Street Journal,* 7 September 1977; Cranfield, "China Gearing Up," p. 22. Chart represents rig status as of mid-1977. More recently, China has reportedly contracted with National Supply Company of Houston for two self-contained rigs for installation on fixed, Chinese-made platforms. Total value of the transaction is estimated at $20 to 30 million, with

delivery scheduled in mid- to late-1978. *Washington Post,* 27 November 1977; *Oil and Gas Journal,* 12 December 1977, p. 35.

51. CIA, *China's Oil Potential,* p. 13; Meyerhoff and Willums, *(CCOP) Technical Bulletin* 10:190.

52. Interview with Raymond Cox, president of Geo Space Corporation, Houston, Texas, 4 February 1976. The two indigenous survey crews operated with analogue pulse-width modulation devices copied from the French, who had in turn adapted the equipment from Exxon (Carter Oil).

53. *Wall Street Journal,* 9 September 1977. The rig from Norway was reportedly obtained for $27 million. Previously protracted, but ultimately unsucessful, negotiations for similar equipment involved Robin Loh Shipyard in Singapore and the Blohm & Voss firm in West Germany.

54. See Jan-Olaf Willums, "China's Offshore Petroleum," *The China Business Review* 4(4):14 (July-August 1977). Based on his Ph.D. research, Willums portrays eight possible options for China's use of foreign technology in developing its offshore reserves. They range from a "Chinese only" policy to full-scale importation of sophisticated Western equipment and technical services. After computer-matching these options against China's assumed goals for resource development (e.g., to minimize foreign dependence, to encourage Chinese economic development by maximizing medium-term oil production, and to improve the PRC's technological competence and efficiency) and examining the likely discovery rates and costs associated with each option, Willums concludes that a "selected equipment import, extensive know-how exchange" strategy is by far the most attractive choice for China's decision makers.

55. Western sources have quoted Deputy Premier Li Hsien-nien to the effect that "offshore production costs us five times as much as onshore production."

56. Smith, "The Next Oil Giant?" p. 37.

57. For a contrary view, see Willums, "China's Offshore

Petroleum," pp. 7-9, 14. Willums sees 1985 development levels which range from a low of 847 offshore wells producing 1 million bpd to a high of "several times" that figure (2,875 wells producing 5 to 6 million bpd). His most likely scenario, using a middle-of-the-road technology option, forecasts some 2,000 wells producing 2.8 to 3.0 million bpd by the early 1980s. These predictions, however, rest on presumptions of the Chinese immediately absorbing the most sophisticated Western technology, of China's equaling or bettering the exploration, planning, and drilling rates of major oil companies, and of near-term annual production increases for offshore areas which are two to three times those achieved by Ta-ch'ing in its prime.

58. Lead time derives in part from previously listed cost and technology factors, but it is more importantly an indicator of the absolute physical limitations on offshore construction, regardless of financial and other resources invested.

59. In addition, several of the most optimistic statements about PRC potential by Japanese officials and businessmen were made in late 1973 and early 1974, a time when Japan's leaders were desperately striving to ease public concern about the security of their nation's crude oil supply.

60. See, for example, K.C. Yeh, *Communist China's Petroleum Situation,* Rand Memorandum, RM-3160-PR (Santa Monica: The Rand Corporation, 1962), p. 1; and Chu-yuan Cheng, *United States Export Potential of Petroleum Equipment to the People's Republic of China,* a report prepared for Office of East-West Trade Analysis, Bureau of East-West Trade, Domestic and International Business Administration, U.S. Department of Commerce, September 1974, Contract No. 4-36289.

61. U.S., Congress, Joint Economic Committee, a compendium of papers on *China: A Reassessment of the Economy,* article by Nai-Ruenn Chen, "An Assessment of Chinese Economic Data: Availability, Reliability and Usa-

bility," 94th Cong., 1st sess., 10 July 1975, pp. 67-68.

62. CIA, *Energy Balance Projections,* pp. 14, 31-32. The CIA energy consumption projection is consistent with other energy usage forecasts keyed primarily to future growth rates in PRC industrial production. See Thomas G. Rawski, "China and Japan in the World Energy Economy," in *The Energy Question: An International Failure of Policy,* ed. E. W. Erickson and L. Waverman (Toronto: University of Toronto Press, 1974) 1:105-107; and V. Smil, "Energy in China: Achievements and Prospects," *China Quarterly* 65:80-81 (March 1976).

63. For a contrary view, see Peter W. Colm, Rosemary Hayes, and Edwin Jones, *Implications of Prospective Chinese Petroleum Developments to 1980,* IDA paper, P-1229 (Arlington, Va.: Institute for Defense Analyses, July 1976), pp. viii, 51-63. IDA believes that the CIA and others have neglected past structural changes in the Chinese economy, causing them to overstate rates of industrial energy consumption and to understate the amount of residential/commercial energy use. IDA sees these changes as:

a. China's rapid transition from a purely agrarian economy, which demanded unusually large investments in heavy industry, thus stressing energy-intensive activities
b. The initial abundance of non-commercial fuels which, once replaced by commercial sources, created a one-time increase in energy consumption unrelated to any increase in economic activity
c. China's likely future changeover to a more energy-efficient capital stock

The institute therefore believes that *overall* energy growth was inflated by assigning more energy requirements to a fast-growing industrial sector than to the slower-growing residential sector. Regardless of the validity of these postulates, it is interesting that IDA still calculates an annual oil consumption increase of 17.5% from 1976 to 1980, a 1980 oil share of total energy use of 32%, and a 1980 absolute oil consumption figure of 2.6 million bpd. Since these figures are consistent with the CIA oil sector projections for 1980, they tend to confirm that near-term petroleum demand will continue to rise at a substantial pace.

64. *The People's Republic of China: A New Era in Sino-American Trade?* (Los Angeles: NAE Research Associates, Inc., December 1975), p. 45. Hua was at that time a deputy premier with responsibilities primarily in agriculture and public security. Conference references to "basic" mechanization apparently equated to roughly 70% mechanization of certain principal tasks by 1980.

65. For example, rice transplanters and power threshers enable Chinese workers to accelerate the transition from the first to second crop without increasing labor requirements. Hand tractors not only assist in the fields but also free labor for the large amounts of construction work involved in China's agricultural transformation. Water is increasingly moved from irrigation canals to fields via power pumps instead of human- or animal-run water wheels. See U.S., Congress, Joint Economic Committee, a compendium of papers on *China: A Reassessment of the Economy,* article by D. H. Perkins, "Constraints Influencing China's Agricultural Performance," 94th Cong., 1st sess., 10 July 1975, p. 355.

66. Allen S. Whiting and Robert F. Dernberger, *China's Future: Foreign Policy and Economic Development in the Post-Mao Era* (New York: McGraw-Hill Book Co., 1977), p. 112.

67. Smil, "Energy in China," pp. 76-77. Smil examines the relatively slow growth in Chinese food production from 1952 to 1972 and concludes that future boosts in agricultural output will require vastly increased fossil fuel subsidies.

68. While improvements in China's efficiency of energy use might slow this rising consumption trend, it seems likely that such improvements will be offset by (a) rapid expansion of industries which consume large amounts of energy per unit of final output (e.g., chemical fertilizers and petrochemicals); and (b) the increasing substitution of commercial for non-commercial fuels in rural areas (e.g., coal's replacement of wood in rural family cooking). See CIA, *Energy Balance Projections,* p. 16. This conclusion is also consistent with increasing Soviet rates of use from the early 1950s (point where China was in early 1970s) to 1960 (consumption level which the PRC is expected to reach in 1980). See CIA, *Soviet Long-Range Energy Forecasts,* A (ER), 75-71, September 1975, pp. 25, 26.

69. China's annual production of diesel locomotives, for instance, has risen from 30 in 1965 to 275 in 1975. See *The China Business Review* 4(2):35 (March-April 1977).

70. CIA, *Energy Balance Projections,* pp. 12, 33.

71. CIA figures on the relative percentages of coal and oil are consistent with other sources. See K. C. Yeh and Yuan-li Wu, "Oil and Strategy" (paper presented to Fifth Sino-American Conference on Mainland China, Taipei, 10 June 1976), p. 7.

72. CIA, *Energy Balance Projections,* pp. 6, 33. Although most of this substitution is due to oil consumption rising much faster than coal usage, there has also been growing evidence of industries in cities such as Peking and Tientsin actually switching from coal to oil or gas. See *New China News Agency,* 24 September 1975, and *Commercial Daily* (Hong Kong), 6-8 October 1975. In addition, European countries have reported that China is modifying many existing coal-fired power plants to burn oil and that the PRC will probably decide to put much of its new generating capacity on a dual-fired basis as well. *World Energy Outlook,* a report by the secretary-general, Organization of Economic Cooperation and Development, 1977, p. 82.

73. China seems to recognize the implications of these oil usage trends. The government sponsored two major conferences in early 1977 to kick off energy conservation campaigns for the industrial and transportation sectors. *The China Business Review* 4(2):49-50 (March-April 1977).

74. CIA, *Energy Balance Projections*, pp. 2, 3, 28.

75. Although recent coal output estimates vary, most sources figure 9 to 11% growth in 1975 and no growth in 1976. See the *Washington Post*, 29 September 1976; *Economic Reporter* (Hong Kong), 28 July 1976; *New China News Agency*, 12 July 1976. The 1975 (and early 1976) increases apparently represented successful labor mobilization efforts before and after the October 1975 National Coal Conference. The 1976 figures remain obscure as the PRC has not released any percentage changes. Most U.S. government sources believe this means little or no increase due to a number of factors: the T'ang-shan earthquake, widespread labor unrest accompanying the "Gang of Four" activities, and the persistence of systemic industry problems.

The impact of the T'ang-shan quake alone was considerable. It completely destroyed the PRC's Kailuan mine, which produced roughly 6% (25 million tons) of all China's coal in 1975 and over 8% of China's large mine (higher quality) production in that year. To separate its effect on PRC coal growth from others is not as yet possible, but suffice it to say that this chance occurrence will only further magnify the industry's existing problems.

76. Even in the 1950s, PRC statistics indicated that fixed assets for industrial production per productive worker in the petroleum industry were five times those in the coal industry. Nai-Ruenn Chen, *Chinese Economic Statistics, A Handbook for Mainland China* (Chicago: Aldine, 1967), pp. 150-151.

77. This situation is further exacerbated by the reported lack of Chinese coal which can be readily beneficiated. (A 1959 Soviet study found that 81% of Chinese coal was either

"difficult or very difficult" to beneficiate.) If correct, this condition will require China to make massive investments in the acquisition of beneficiation technology. Cited in U.S., Congress, Joint Economic Committee, a compendium of papers on *China: A Reassessment of the Economy,* article by Alfred H. Usack, Jr. and James D. Egan, "China's Iron and Steel Industry," 94th Cong., 1st sess., 10 July 1975, p. 270.

78. From 1957 to 1974, small mines' share of total coal production rose from 6 to 28%. CIA, *Energy Balance Projections,* p. 4.

79. A. B. Ikonmikov, *Mineral Resources of China* (Boulder, Co.: The Geological Society of America, 1975), p. 268.

80. See Appendix A; CIA, *China: Oil Production Prospects,* p. 9; *Oil Daily,* 6 January 1977, p. 1; and *New China News Agency,* 26 December 1977. Figure for 1977 estimated on the basis of 8 percent growth rate announced by the PRC for first eleven months.

81. CIA, *Energy Balance Projections,* pp. 11-16.

82. Ibid., p. 16. An energy-GNP coefficient of 1.45 instead of 1.42 is assumed due to rapid expansion of industries which consume large amounts of energy per final unit of output. Although the result is an export range of 540,000 to 660,000 bpd (midpoint: 600,000), CIA has recently revised this figure downward to a more likely 1980 export level of 500,000 bpd. See U.S., CIA, *The International Energy Situation: Outlook to 1985,* ER 77-102404, April 1977, p. 5. Both the 500,000 and 600,000 bpd export levels are consistent with other near-term supply/demand forecasts. See Richard M. Evans, "China and the Energy Crisis" (paper for Center of International Affairs, Harvard University, March 1975), pp. 25-26; Rawski, "China and Japan," pp. 105-109; and Smil, "Exports Revisited," p. 73. Smil's work in particular is interesting; it uses a long-range GNP-energy linear regression equation, instead of the CIA's energy-GNP elasticity coefficient approach, to reach approximately the same

energy surplus figures as CIA for 1980 and 1985.

83. CIA, *Energy Balance Projections,* p. 1; CIA, *International Energy Situation,* p. 15. Differences between both numbers reflect CIA's revision of their forecast between publication of the first (November 1975) and second (April 1977) references. This is perhaps a good example of the limitations of econometric energy studies referred to in the next paragraph of the text.

84. JEC, Usack and Egan, "China's Iron and Steel Industry," p. 284. In addition, 1976 steel output reportedly fell to 21 million tons from a 1975 level of 26 million tons. *The China Business Review* 4(2):49 (March-April 1977).

85. For a more detailed analysis of this phenomenon, see Whiting and Dernberger, *China's Future,* pp. 98-101.

86. The North China Plain, which contains one-fifth of China's population, is characterized by dryland wheat farming and inadequate water supply. To get the full benefit of using chemical fertilizers under these conditions, their application must go hand-in-hand with increased irrigation.

87. In the wake of Hua Kuo-feng's purge of the "Gang of Four" and their supporters, the new Chinese leadership is rewriting the current Five Year Plan. Although this process has been complicated by a breakdown in management and accounting procedures during the factional struggles of 1976, the new plan is sufficiently well-developed to indicate the Chinese are apparently considering a higher investment priority for coal than for oil in the near future. See CIA, *China: Oil Production Prospects,* p. 23.

88. A. A. Meyerhoff, "Chinese Petroleum Industry Potential" (working group from California Arms Control and Foreign Policy seminar on U.S. Relations in the Pacific, 24 May 1974), p. 5. A 4 million bpd level in 1980 would be the result of a 20% annual growth rate to that year. Meyerhoff sees the main costs coming from storage facilities, pipelines, compressors, and increased refinery capacity.

89. Cheng, *China's Petroleum Industry*, pp. 189-190. Cheng actually estimates a total requirement for $45 billion ($4.5 billion a year) to reach his predicted output level of 6.7 million bpd in 1985. Applying the same capital investment/output ratio to production levels of 4 million bpd in 1980 and 6 million bpd in 1985 yields yearly investment figures of $3.8 and $3.6 billion, respectively.

90. The methodology and 1953-1957 investment figures which follow come from Williams, "The Chinese Petroleum Industry," pp. 248-249, although base periods for calculation have been changed to ensure consistency. GNP figures are derived and projected from U.S., Congress, JEC compendium on *China: A Reassessment of the Economy*, article by Arthur G. Ashbrook, Jr., "China: An Economic Overview, 1975," 94th Cong., 1st sess., 10 July 1975, pp. 42-44.

91. U.S., Congress, JEC, compendium on *China: A Reassessment of the Economy*, article by Hans Heymann, Jr., "Acquisition and Diffusion of Technology in China," 94th Cong., 1st sess., 10 July 1975, p. 692.

92. Lack of sufficient geologists and geophysicists, for example, could hamper exploration activities by (a) not giving seismic teams adequate general guidance on what kinds of structures to map and analyze; and (b) limiting survey crew capacity to identify specific parameters indicating oil presence in different geologic formations.

93. Interview with Raymond Cox, Geo Space, on 4 February 1976. Other sources cite lack of specialized manufacturing skills and quality control techniques as the deficiencies in major plant (e.g., production platforms, refineries, petrochemical facilities) operations. See Ling, *Petroleum Industry of the PRC*, p. 82.

94. For further details, see R. F. Dernberger, "The Transfer of Technology to China," *Asia Quarterly* 3:239-242 (1974). Dernberger cites estimates that China's official rate of economic growth from 1953 to 1957 would have been 20 to 30

percent lower without foreign equipment imports.

95. To illustrate the gap to be bridged, some knowledgeable observers believe PRC mechanical production technology lags 20 years behind the West. Robert G. Chollar, president of Charles F. Kettering Foundation, "The China Model: Some Observations on Its Vulnerability to Change" (report from trip of private sector foreign policy leaders to the PRC, 6-23 October 1975), p. 5.

96. As of August 1977, Hua's main allies, veteran bureaucrats and the military, comprised roughly three-fourths of the new Chinese Communist Party Central Committee. Over one-half of the people dropped from full committee membership and two-thirds of those released as alternate members were "farm and factory activists." *Washington Post,* 2 September 1977.

97. The Long March took place from 1934 to 1936 and involved the relocation of a Mao-led Chinese Communist Party (CCP) from southwest to north central China. Pursued by overwhelming Chinese Nationalist forces, Mao guided the CCP over 2,000 miles under the most arduous of conditions to establish a secure headquarters in the caves of Yennan. The March has become an epic event in CCP history and has served in the past as sort of a bench mark for measurement of one's revolutionary credentials.

98. The terms used for classification, as well as much of the description of each group, come from Michael Oksenberg and Steven Goldstein, "The Chinese Political Spectrum," *Problems of Communism* 23(2):1-13 (March-April 1974); and Peter G. Mueller and Douglas A. Ross, *China and Japan—Emerging Global Powers* (New York: Praeger Publishers, 1975), pp. 163-173.

99. *The China Business Review* 4(4):29 (July-August 1977).

100. In addition, Indonesia has had the advantage of joint ventures and like forms of close cooperation with the major oil companies which China is not willing to accept.

101. See *The China Business Review* 4(3):7 (May-June 1977).

102. Interview with Raymond Cox, Geo Space, 4 February 1976. Although this is the only apparent instance of oil-related training in the U.S. to date, approximately 2,000 Chinese in other fields have received similar host country instruction over the past five years.

103. *Wall Street Journal,* 31 August 1977, p. 6. Although Chinese leaders refuse to look on supplier credits as debt, they employ a variety of deficit-type financing mechanisms:

a. Foreign banks are permitted to finance exports to the PRC indirectly, often by supporting the credit offered by companies selling to China.
b. The Bank of China regularly operates in the interbank markets of Europe and Hong Kong. It has recently been a net borrower of Eurodollars due in one to three years.
c. The Bank of China maintains offsetting deposits with several foreign banks, especially those in Japan. These banks deposit yen or similar hard currency in Chinese banks in exchange for PRC deposits of renminbi, China's national currency. Since the renminbi is not convertible, the offsetting deposits amount to hard currency loans to Peking.
d. The Bank of China, although avoiding the practice to date, could easily obtain syndicated loans or credit from a group of foreign banks on the interbank market.

104. See, for example, *A New Era in Sino-American Trade?,* p. 95.

105. Much of the self-reliance description draws on Heymann, "Acquisition of Diffusion," pp. 689-691.

106. For an example of a specific PRC policy statement, see Teng Hsiao-peng's April 1974 speech to the United

Nations General Assembly (UNGA). In the words of a Houston oil executive: "The Chinese want to go it alone. They are not interested in the profit-sharing, the production-sharing, or the joint ventures that are common between oil companies and other foreign governments. They don't want foreigners in their country and definitely don't want us there as an investor." *U.S. News and World Report,* 9 December 1975. There have been some reports, however, of a less unequivocal PRC attitude toward oil companies in Australia and Italy.

107. See interviews with Vice-Premier Li Hsien-nien and Vice Foreign Minister Yu Chun, *Wall Street Journal,* 3 October 1977 and 4 October 1977.

108. Conversely, after defining the full extent of its reserves and building up its technological base in presently deficient areas, the PRC could phase out even limited foreign assistance activities and revert back to direct equipment purchases as its principal means of technology transfer.

109. For example, the PRC sold oil at "friendship prices" (between $7.50 and $10 a barrel) to both Thailand and the Philippines before either had established diplomatic relations with Peking. Since this was low-quality Sheng-li crude, however, it is difficult to separate the political and economic motivations behind China's actions.

110. Chart represents a composite from the following sources: *A New Era in Sino-American Trade?,* p. 25; *Platt's Oilgram Service,* 28 January 1976, p. 1; *Wall Street Journal,* 1 April 1976 and 20 September 1977.

111. Sakhalin offshore activity appears to be one of the genuine bright spots in present USSR-Japanese natural resource endeavors. Four test wells have yielded 7,000 bpd of high-quality crude (0.25% sulfur, 1.26% paraffin) and predictions of large potential reserves. Still, full assessment of the field's commercial promise is four to five years away. *New York Times,* 15 October 1977, p. 63.

112. For further elaboration, see Chalmers Johnson, "How China and Japan View Each Other," *Foreign Affairs* 50(4):711-712 (July 1972).

113. The late 1977 price was $13.20 per barrel, f.o.b.

114. *Wall Street Journal,* 1 December 1977 and 17 February 1978.

115. For example, Chinese Foreign Trade Minister Li Chiang has been one of the main advocates of a long-term trade pact under which China would export oil and coal to Japan in exchange for industrial plant and technology imports. Although this is probably indicative of favorable Chinese economic attitudes in the post-Mao era, near-term oil exports still appear constrained by factors in both countries. On the Japanese side, demand for Ta-ch'ing crude is limited to requirements for its direct burning in power plants, to those quantities which can be blended with lighter crudes, and to products from those few refineries which, as described later in the text, can convert to handle it in the first place. With regard to the PRC, persistent suspicions exist that Ta-ch'ing may never be able to sustain exports in excess of 600,000 bpd; they tend to be supported by China's claims that near-term exports to Japan will not exceed 200,000 bpd. *The China Business Review* 4(3):53 (May-June 1977) and 4(4):32-33 (July-August 1977); *Platt's Oilgram Service,* 11 February 1977, p. 3; *Baltimore Sun,* 30 September 1977. The last reference quotes members of a Japanese Diet (parliament) delegation that visited China in August 1977, to the effect that crude export prospects were limited to 200,000 bpd for the immediate future but that a level of 400,000 bpd was possible if Japanese importers could provide the necessary purifying devices *to PRC refineries.* While normally not attractive from an economic standpoint, petroleum *product* exports from China to Japan could offer environmental and supply security advantages. In any case, it seems to be another possible option which both countries will explore.

116. To date only Idemitsu Kosan has decided to add such

dewaxing capacity at its refinery in Hyogo County. The installation would involve expanding the existing facility at that location from 110,000 to 210,000 bpd; it would probably cost $150 to $200 million and would take roughly three years to complete. Additionally, MITI issued a directive in August 1977 *requiring* that Japanese refiners process low grade crudes such as Ta'ch'ing oil. Pursuant to that action, the quasi-governmental Japan Petroleum Development Corporation and several refining companies apparently plan to establish a joint corporation to invest in construction of such specialized refining facilities. Although the government will pay some of the costs, its precise role—and therefore the likely speed with which the dewaxing problem will be resolved—remains unclear. It is also not clear how Idemitsu's prior plans would fit into this picture. *The China Business Review* 4(3):53 (May-June 1977) and 4(5):51 (September-October 1977).

117. Such a switch apparently occurred for a brief period in mid-1976; PRC negotiators with Japan abruptly switched from the attitude that oil exports were to be encouraged to a view which placed higher priority on fulfilling domestic needs. Elimination of the "Gang of Four" now seems to have returned China to its former pro-export outlook. See *China: Oil Production Prospects,* p. 23.

118. See, for example, a 1976 study by Japan's Institute of Energy Economics which predicts an annual oil growth of 6.4% until 1985, with total imports from the PRC reaching 860,000 bpd in that year. *Platt's Oilgram Service,* 30 December 1976, p. 2.

119. Both nations are presently taking low-key approaches—China because it needs Japan's near-term cooperation and does not currently have the capability or the need to drill near the Senkakus; Japan because it hopes that the PRC will eventually turn its way for offshore development assistance.

120. A good example of this mutual limitation pheno-

menon, albeit in a somewhat different context, is North Korea. When either the Soviets or the Chinese deviate even slightly from the North's hard-line hostility to the South—as with Soviet admission of an occasional South Korean to Moscow, or China's "compromise" proposal on Korea during the 1973 UNGA session—the rival seeks to exploit the other's action to gain influence with North Korea.

121. Both general and country-specific themes were perhaps best illustrated in Chou En-lai's January 1975 speech to the Fourth National People's Congress.

122. These concerns relate mainly to Vietnam's relatively pro-Soviet stance, possible USSR base rights at Camranh Bay, and the growing Far East/Indian Ocean presence of the Soviet navy. See *FEER,* 30 January 1976, p. 9.

123. For example, see Teng Hsiao-peng's April 1974 speech to the Sixth Special Session of the UNGA.

124. For a more detailed discussion, see *A New Era in Sino-American Trade?,* pp. 22-23, 153-154, 160.

125. As shown in Appendix V, imports are expected to rise from 4.76 million bpd in 1976 to 5.68 million bpd in 1980. If a slightly higher 6% annual import growth rate is used, 1980 imports will equal 6.01 million bpd. In either case, a 1 million bpd import level should only represent about 17% of the total Japanese requirements.

126. Iain F. Elliot, *The Soviet Energy Balance* (New York: Praeger Publishers, Inc., 1974), pp. 76-77; and Robert I. Ebel, *Communist Trade in Oil and Gas* (New York: Praeger Publishers, Inc., 1970), p. 40.

127. See, for example, Cheng, *China's Petroleum Industry,* p. 39.

128. Ibid. The USSR produced 45 million tons of steel, 170 billion kilowatt hours of electricity, and 731,000 tons of oil production and refinery equipment in 1954/1955. Comparable Chinese figures were 25 million tons of steel (1973), 100 to 125 billion kilowatt hours of electricity (1971),

and an undetermined but substantially lower level of oil-associated machinery.

129. Heyman, "Acquisition and Diffusion of Technology," p. 689. Using Heymann's calculation that net fixed capital formation represents about 16% of GNP, increases in petroleum percentages of net capital investment from 3.0% in 1973 to 5.6% in 1978, or from 3.9% in 1975 to 7.4% in 1980, indicate *additional* resource allocations to the oil sector of between $1.2 and $1.8 billion per year. Reprogramming $1.8 billion in a total capital pool of $51.6 billion (16% of 1980 GNP) would appear to present considerable problems.

130. It is realized, of course, that the value of any joint probability depends on the number of discrete factors which affect the outcome. But since this paper has hopefully demonstrated both that each factor listed is essential to success of the overall event (i.e., producing 8 million bpd by 1989/1990), and that each is statistically independent of the others, the overall joint probability is rigorous. It is emphasized, however, that these probabilities are subjective values assigned by the author, based on a "moderately optimistic" assessment of the various qualitative considerations discussed in this study.

131. The sources listed below have made the following estimates of out-year PRC crude oil surpluses:

Source	PRC Export	Availability	(Million bpd)
	1980	1985	1990
1	0.5	0	--
2	0.5 - 0.6	1.2	--
3	0.5	1.0	0
4	---	0.3 - 0.5	0

1. CIA, *International Energy Situation,* p. 15.
2. OECD, *World Energy Outlook,* p. 82.
3. *Project Interdependence: United States and World*

Energy Outlook Through 1980, prepared for the Committee on Energy and Natural Resources; the National Ocean Policy Study of the Committee on Commerce, Science, and Transportation, U.S. Senate and Subcommittee on Energy and Power, Committee on Interstate and Foreign Commerce, U.S. House of Representatives, by the Congressional Research Service, Library of Congress, June 1977, p. 73.

4. James A. West, associate assistant administrator, Office of International Energy Affairs, Federal Energy Administration, "The Pacific Area Energy Supply/Demand Outlook—The Next 15 Years" (speech to the Second Pacific Chemical Engineering Congress, Denver, Colo., August 1977), p. 3.

132. See Peter Weintraub, "The New Shopping List," *FEER,* 7 October 1977, p. 51.

133. CIA, *Energy Balance Projections,* p. 1. Again, this was CIA's 1975 supply/demand study, the results of which were revised downward by release of the *International Energy Situation* and *China: Oil Production Prospects* in April and June 1977, respectively.